T0214638

Lecture Notes in Computer Science   11452

Commenced Publication in 1973
Founding and Former Series Editors:
Gerhard Goos, Juris Hartmanis, and Jan van Leeuwen

More information about this series at http://www.springer.com/series/7407

Arnaud Liefooghe · Luís Paquete (Eds.)

Evolutionary Computation in Combinatorial Optimization

19th European Conference, EvoCOP 2019
Held as Part of EvoStar 2019
Leipzig, Germany, April 24–26, 2019
Proceedings

 Springer

Editors
Arnaud Liefooghe (iD)
University of Lille
Lille, France

Luís Paquete
University of Coimbra
Coimbra, Portugal

ISSN 0302-9743   ISSN 1611-3349 (electronic)
Lecture Notes in Computer Science
ISBN 978-3-030-16710-3   ISBN 978-3-030-16711-0 (eBook)
https://doi.org/10.1007/978-3-030-16711-0

Library of Congress Control Number: 2019936143

LNCS Sublibrary: SL1 – Theoretical Computer Science and General Issues

This Springer imprint is published by the registered company Springer Nature Switzerland AG
The registered company address is: Gewerbestrasse 11, 6330 Cham, Switzerland

Preface

Combinatorial optimization problems arise very often in the demanding life that each of us face everyday, from timetabling to planning routes, from resource allocation to the location of facilities. The main challenge to the scientific community is to develop solution methods that are able to achieve high-quality solutions to these problems in a short amount of time. This is, by far, a non-trivial task. The underlying difficulty of these problems poses relevant research questions concerning the trade-off between solution quality and run-time that can ever be achieved on these problems. Evolutionary techniques and related metaheuristics are particularly well suited for grasping these trade-offs since they are able to return solutions at *anytime* of their search process. Their search strategies are simple and intuitive, very much tied to human reasoning for problem solving. For this reason, they have been extremely popular among both the scientific community and practitioners, and they are selected as de facto techniques for solving real-life optimization problems.

This volume contains the proceedings of EvoCOP 2019, the 19th European Conference on Evolutionary Computation in Combinatorial Optimization. The conference was held in Leipzig, Germany, during April 24–26, 2019. The EvoCOP conference series started in 2001, with the first workshop specifically devoted to evolutionary computation in combinatorial optimization. It became an annual conference in 2004. EvoCOP 2019 was organized together with EuroGP (the 22nd European Conference on Genetic Programming), EvoMUSART (the 8th International Conference on Computational Intelligence in Music, Sound, Art and Design), and EvoApplications (the 22nd European Conference on the Applications of Evolutionary Computation, formerly known as EvoWorkshops), in a joint event collectively known as EvoStar 2019. Previous EvoCOP proceedings were published by Springer in the *Lecture Notes in Computer Science* series (LNCS volumes 2037, 2279, 2611, 3004, 3448, 3906, 4446, 4972, 5482, 6022, 6622, 7245, 7832, 8600, 9026, 9595, 10197, and 10782). The table on the next page reports the statistics for each of the previous conference.

This year, 13 out of 37 papers were accepted after a rigorous double-blind process, resulting in a 35% acceptance rate. We would like to acknowledge the quality and timeliness of our Program Committee members' work. Decisions considered both the reviewers' report and the evaluation of the program chairs. The 14 accepted papers cover a wide spectrum of topics, ranging from the foundations of evolutionary computation algorithms and other search heuristics, to their accurate design and application to combinatorial optimization problems. Fundamental and methodological aspects deal with runtime analysis, the structural properties of fitness landscapes, the study of metaheuristics core components, the clever design of their search principles, and their careful selection and configuration. Applications cover problem domains such as scheduling, routing, partitioning, and general graph problems. It is our hope that the range of topics covered in this volume of EvoCOP proceedings reflects the current state of research in the fields of evolutionary computation and combinatorial optimisation.

EvoCOP	LNCS vol.	Submitted	Accepted	Acceptance (%)
2019	11452	37	13	35.1
2018	10782	37	12	32.4
2017	10197	39	16	41.0
2016	9595	44	17	38.6
2015	9026	46	19	41.3
2014	8600	42	20	47.6
2013	7832	50	23	46.0
2012	7245	48	22	45.8
2011	6622	42	22	52.4
2010	6022	69	24	34.8
2009	5482	53	21	39.6
2008	4972	69	24	34.8
2007	4446	81	21	25.9
2006	3906	77	24	31.2
2005	3448	66	24	36.4
2004	3004	86	23	26.7
2003	2611	39	19	48.7
2002	2279	32	18	56.3
2001	2037	31	23	74.2

We would like to express our appreciation to the various persons and institutions who helped make EvoCOP 2019 a successful event. Firstly, we thank the local organization team led by Hendrik Richter, the Faculty of Electrical Engineering and Information Technology of HTWK Leipzig University of Applied Sciences for supporting the local organization, and the HTWK Leipzig University of Applied Sciences for their patronage of the event. We extend our acknowledgments to Marc Schoenauer from Inria Saclay, France, for his continued assistance in providing the MyReview conference management system, and to Pablo García-Sánchez from the University of Cádiz, Spain, for taking care of the EvoStar publicity, website, and social media. Thanks are also due to our EvoStar coordinator Anna I Esparcia-Alcázar, from Universitat Politècnica de València, Spain, as well as to the SPECIES (Society for the Promotion of Evolutionary Computation in Europe and its Surroundings) executive board, including Marc Schoenauer (President), Anna I Esparcia-Alcázar (Secretary and Vice President), and Wolfgang Banzhaf (Treasurer). We finally wish to thank our prominent keynote speakers, Manja Marz from the Friedrich Schiller University Jena, Germany, and Risto Miikkulainen from the University of Texas, USA.

Special thanks also to Christian Blum, Francisco Chicano, Carlos Cotta, Peter Cowling, Jens Gottlieb, Jin-Kao Hao, Bin Hu, Jano van Hemert, Manuel Lopéz-Ibáñez, Peter Merz, Martin Middendorf, Gabriela Ochoa, and Günther R. Raidl for their hard work and dedication at past editions of EvoCOP, making this one of the reference international events in evolutionary computation and metaheuristics.

April 2019   Arnaud Liefooghe
                                                              Luís Paquete

Organization

EvoCOP 2019 was organized as a part of EvoStar 2019, jointly with EuroGP 2019, EvoMUSART 2019, and EvoApplications 2019.

Organizing Committee

Program Chairs

Arnaud Liefooghe   University of Lille, France
Luís Paquete   University of Coimbra, Portugal

Local Organization

Hendrik Richter   Leipzig University of Applied Sciences, Germany

Publicity Chair

Pablo García-Sánchez   University of Cádiz, Spain

EvoCOP Steering Committee

Christian Blum   Artificial Intelligence Research Institute (IIIA-CSIC),
                            Bellaterra, Spain
Francisco Chicano   University of Málaga, Spain
Carlos Cotta   University of Málaga, Spain
Peter Cowling   University of York, UK
Jens Gottlieb   SAP AG, Germany
Jin-Kao Hao   University of Angers, France
Bin Hu   AIT Austrian Institute of Technology, Austria
Jano van Hemert   Optos, UK
Manuel Lopéz-Ibáñez   University of Manchester, UK
Peter Merz   Hannover University of Applied Sciences and Arts,
                            Germany
Martin Middendorf   University of Leipzig, Germany
Gabriela Ochoa   University of Stirling, UK
Günther Raidl   Vienna University of Technology, Austria

Society for the Promotion of Evolutionary Computation in Europe and Its Surroundings (SPECIES)

Marc Schoenauer (President)
Anna I Esparcia-Alcázar (Secretary and Vice President)
Wolfgang Banzhaf (Treasurer)

Program Committee

Hernán Aguirre	Shinshu University, Japan
Enrique Alba	University of Málaga, Spain
Richard Allmendinger	University of Manchester, UK
Thomas Bartz-Beielstein	Cologne University of Applied Sciences, Germany
Matthieu Basseur	University of Angers, France
Benjamin Biesinger	Austrian Institute of Technology, Austria
Christian Blum	Artificial Intelligence Research Institute (IIIA-CSIC), Bellaterra, Spain
Sandy Brownlee	University of Stirling, UK
Pedro Castillo	University of Granada, Spain
Marco Chiarandini	University of Southern Denmark, Denmark
Francisco Chicano	University of Málaga, Spain
Carlos Coello Coello	CINVESTAV-IPN, Mexico
Carlos Cotta	University of Málaga, Spain
Bilel Derbel	University of Lille, France
Karl Doerner	Johannes Kepler University Linz, Austria
Carola Doerr	CNRS and Sorbonne University Paris, France
Benjamin Doerr	LIX-Ecole Polytechnique, France
Paola Festa	Universitá di Napoli Federico II, Italy
Carlos M. Fonseca	University of Coimbra, Portugal
Bernd Freisleben	University of Marburg, Germany
Carlos Garcia-Martinez	University of Córdoba, Spain
Adrien Goeffon	University of Angers, France
Jens Gottlieb	SAP SE, Germany
Andreia Guerreiro	University of Coimbra, Portugal
Walter Gutjahr	University of Vienna, Austria
Jin-Kao Hao	University of Angers, France
Emma Hart	Edinburgh Napier University, UK
Geir Hasle	SINTEF Digital, Norway
Bin Hu	Austrian Institute of Technology, Austria
Thomas Jansen	Aberystwyth University, UK
Andrzej Jaszkiewicz	Poznan University of Technology, Poland
Ahmed Kheiri	Lancaster University, UK
Mario Koeppen	Kyushu Institute of Technology, Japan
Timo Kötzing	Hasso Plattner Institute, Germany
Frederic Lardeux	University of Angers, France
Per Kristian Lehre	University of Birmingham, UK
Rhyd Lewis	Cardiff University, UK
Arnaud Liefooghe	University of Lille, France
Manuel López-Ibáñez	University of Manchester, UK
Jose Antonio Lozano	University of the Basque Country, Spain
Gabriel Luque	University of Málaga, Spain
Juan Julian Merelo	University of Granada, Spain
Krzysztof Michalak	University of Economics, Wroclaw, Poland

Martin Middendorf	University of Leipzig, Germany
Christine L. Mumford	Cardiff University, UK
Nysret Musliu	Vienna University of Technology, Austria
Gabriela Ochoa	University of Stirling, UK
Pietro Oliveto	University of Sheffield, UK
Beatrice Ombuki-Berman	Brock University, Canada
Luís Paquete	University of Coimbra, Portugal
Mario Pavone	University of Catania, Italy
Paola Pellegrini	French Institute of Science and Technology for Transport, France
Francisco J. Pereira	University of Coimbra, Portugal
Daniel Porumbel	CNAM, France
Jakob Puchinger	SystemX-Centrale Supélec, France
Günther Raidl	Vienna University of Technology, Austria
Maria Cristina Riff	Universidad Técnica Federico Santa María, Chile
Marcus Ritt	Universidade Federal do Rio Grande do Sul, Brazil
Eduardo Rodriguez-Tello	CINVESTAV - Tamaulipas, Mexico
Andrea Roli	Università di Bologna, Italy
Peter Ross	Edinburgh Napier University, UK
Frederic Saubion	University of Angers, France
Patrick Siarry	University of Paris 12, France
Kevin Sim	Edinburgh Napier University, UK
Jim Smith	University of the West of England, UK
Giovanni Squillero	Politecnico di Torino, Italy
Thomas Stützle	Université Libre de Bruxelles, Belgium
Andrew Sutton	University of Minnesota Duluth, USA
El-ghazali Talbi	University of Lille, France
Sara Tari	University of Angers, France
Renato Tinós	University of São Paulo, Brazil
Nadarajen Veerapen	University of Lille, France
Sebastien Verel	Université du Littoral Cote d'Opale, France
Darrell Whitley	Colorado State University, USA
Carsten Witt	Technical University of Denmark, Denmark
Bing Xue	Victoria University of Wellington, New Zealand
Takeshi Yamada	NTT Communication Science Laboratories, Japan
Christine Zarges	Aberystwyth University, UK

Contents

A Cooperative Optimization Approach for Distributing Service Points in Mobility Applications

Thomas Jatschka[1]([✉]), Tobias Rodemann[2]([✉]), and Günther R. Raidl[1]([✉])

[1] Institute of Logic and Computation, TU Wien, Vienna, Austria
{tjatschk,raidl}@ac.tuwien.ac.at
[2] Honda Research Institute Europe, Offenbach/Main, Germany
tobias.rodemann@honda-ri.de

Abstract. We investigate a variant of the facility location problem concerning the optimal distribution of service points with incomplete information within a certain geographical area. The application scenario is generic in principle, but we have the setup of charging stations for electric vehicles or rental stations for bicycles or cars in mind. When planning such systems, estimating under which conditions which customer demand can be fulfilled is fundamental in order to evaluate and optimize possible solutions. In this paper we present a cooperative optimization approach for distributing service points that incorporates potential customers not only in the data acquisition but also during the optimization process. A surrogate objective function is used to evaluate intermediate solutions during the optimization. The quality of this surrogate objective function is iteratively improved by learning from the feedback of potential users given to candidate solutions. For the actual optimization we consider a population based iterated greedy algorithm. Experiments on artificial benchmark scenarios with idealized simulated user behavior show the learning capabilities of the surrogate objective function and the effectiveness of the optimization.

Keywords: Cooperative optimization · Facility location problem · Surrogate objective function · Metaheuristics

1 Introduction

Identifying optimal locations for setting up charging stations for electric vehicles (EVs), rental stations in public bike or car sharing systems, or, more generally some kind of service stations for mobility applications is always a challenging problem when planning such systems. Usually, the goal is to place stations at locations with high customer demand in order to maximize the usage and revenue

T. Jatschka—acknowledges the financial support from Honda Research Institute Europe.

A. Liefooghe and L. Paquete (Eds.): EvoCOP 2019, LNCS 11452, pp. 1–16, 2019.
https://doi.org/10.1007/978-3-030-16711-0_1

of such systems. However, estimating the customer demand that can possibly be fulfilled is challenging. Demographic data is usually interlinked with geographic information, data on public transport, the street network, knowledge on manifold special locations, etc. Additionally, surveys of potential customers are performed. Customer demand information determined in such ways typically is vague, and not uncommonly a system built on such assumptions is not as effective as originally hoped for due to major deviations in reality. The actual usage of a service system by a user will in general depend not only on the construction of service points on a few specific locations but more globally on non-trivial relationships of the user's necessities and preferences in conjunction with larger parts of the whole service system. For example in the case of charging stations for EVs, consider the situation that a station is not built in close proximity to a location a user is interested in, e.g., the user's place of work. Intuitively, one might say that the user's demand cannot be fulfilled. However, such a conclusion may be too naive. It might easily be the case that some other location is covered that provides a reasonable alternative for the user, e.g., by additionally using some public transport. Thus, there might be alternatives for fulfilling demand that cannot all be foreseen or exactly pre-specified by potential users.

Hence, a crucial assumption that makes the situation challenging and appears to be particularly valid in the context of the above mentioned mobility applications is the following: We are in general not able to obtain complete information from potential users about the conditions under which how much demand will be fulfilled, even when assuming absolutely rational users, neglecting uncertainty, and ignoring aspects arising from competition from many users on possibly scarce resources.

To overcome this problem, we investigate a *cooperative optimization approach*. More generally, interactive optimization approaches incorporate potential users on a large scale and more tightly into the data acquisition as well as the optimization process; for a review see [19]. We confront the potential users with carefully selected candidate solutions and ask how these suit the needs. Obtained feedback is used to incrementally gain more knowledge on how much demand may be fulfilled under which conditions. The optimization core relies on a surrogate objective function that approximates the real fulfilled demand. It is based on machine learning models that are trained by the user feedback. Having obtained a new so far best solution from the optimization core, new, more promising candidate solutions can be derived and again be presented to the users. The process is iterated on a large scale with many potential users and several rounds until a satisfactory solution is reached.

We test the approach in a proof-of-concept manner on artificial benchmark scenarios simulating user behavior in an idealized fashion. Results document the learning capabilities of the surrogate objective function and the effectiveness of the optimization.

The paper is structured as follows. In Sect. 2 the Service Point Distribution Problem (SPDP) is formally introduced. Section 3 discusses related work. Section 4 introduces the cooperative optimization approach (COA) for solving the SPDP. Finally, in Sect. 5 we experimentally evaluate the COA and present obtained results.

2 The Service Point Distribution Problem

In this section we specify the problem we consider more formally. In the *Service Point Distribution Problem* (SPDP) we are given a set of locations $V = \{1,\ldots,n\}$ at which service points may be built and a set of potential users $U = \{1\ldots,m\}$. The fixed costs for setting up a service point at location $v \in V$ are $c_v \geq 0$, and this service point's maintenance over a defined time period is supposed to induce variable costs $z_v \geq 0$. The total construction costs must not exceed a maximum budget $B > 0$. Erected service stations may satisfy customer demand, and for each unit of satisfied customer demand a prize $p > 0$ is earned. We remark that for simplicity we do not consider here any capacity restrictions at service points. A solution to the SPDP is given by a binary incidence vector $x = (x_v)_{v \in V}$, where $x_v = 1$ indicates that a service point is to be set up at location v.

The problem is incompletely specified in the sense that we do not have a function for calculating the fulfilled demand for a candidate solution upfront. Instead, we assume here that we are only able to evaluate solutions "exactly" by presenting them to the potential customers U and collecting their feedback. These user evaluations are denoted by $d(u,v,x)$ which specifies the demand of user $u \in U$ fulfilled at location $v \in V$ in solution x. If a service station is not ideal for a user but somewhat acceptable for him to be used and there are no better alternatives in solution x, this is modeled by a correspondingly reduced fulfilled demand value $d(u,v,x)$; i.e., the user is less likely to use this service point and therefore the expected fulfilled demand is lower. Clearly, the number of candidate solutions that are evaluated in this interactive way are a major concern. We cannot confront each user with hundreds or thousands of evaluation requests. Instead, we carefully have to select the solutions to be evaluated by each user in an individual fashion, avoiding redundancies as far as possible.

Naturally, the demand fulfilled at any location must always be non-negative and can only be positive when a service point is set up there, i.e.,

$$d(u,v,x) \geq 0, \quad x_v = 0 \rightarrow d(u,v,x) = 0 \quad u \in U,\ v \in V. \tag{1}$$

A solution x is feasible if its total fixed costs do not exceed the maximum budget B, i.e.,

$$c(x) = \sum_{v \in V} c_v x_v \leq B. \tag{2}$$

The objective is to find a feasible solution that maximizes the prizes earned for satisfied customer demands reduced by the variable costs for maintaining the service points

$$f(x) = p \cdot \sum_{u \in U} \sum_{v \in V} d(u,v,x) - \sum_{v \in V} z_v x_v. \tag{3}$$

3   Related Work

The SPDP can be classified as a variant of the *Facility Location Problem* (FLP). In the FLP a set of potential facility sites and a set of demand points is given. The task is to select a subset of these sites in order to serve the demand points w.r.t. some optimization goal subject to a set of constraints. For a survey on FLPs see [6], for a more comprehensive book on location theory see [11]. More specifically, our SPDP is closely related to the uncapacitated FLP [4] in which each facility can satisfy an arbitrary amount of demand – with the substantial difference that in our case user demands are not known upfront but must be learned via user interaction.

The problem of optimizing the distribution of charging stations for EVs has gained increased attention recently. An essential question of contributions concerning this topic always is how to determine potential customer demands. Chen et al. [2] substitute charging demand with parking demand in order to identify good locations for public charging stations. The parking demand is derived from parking information of a travel survey. In [9], a maximal covering model [3] for identifying charging stations is proposed. The demands are estimated using regression analysis based on surveys on the number of cars per household, the average travel distance of cars, the estimated range of an EV etc. In [1], the charging demand of a location is modeled as the expected duration of charging all drivers that need to charge their EV at this location. The number of drivers in a location is derived from a mobility survey as part of a case study of the city of Coimbra, Portugal. In [14] charging stations for an on-demand bus system are located using taxi probe data of Tokyo. While the focus in this paper lies on distributing charging stations for EVs, our approach is a general framework capable of planning service point based systems of any kind, such as bike or car sharing systems [10].

Opposite to the aforementioned contributions, we assume to have essentially no knowledge on customer demand in advance but aim at obtaining this information on the fly in an interactive way by integrating potential customers in the optimization process. More generally, in *interactive optimization algorithms*, humans are typically used to evaluate the quality of solutions; e.g., in [15] an interactive genetic algorithm for designing dresses is proposed. Instead of explicitly defining a fitness function, the fitness of a solution is decided by a user. For a survey on interactive optimization algorithms see [19]. A major disadvantage of interactive algorithms is that their performance strongly depends on the quality of the feedback given by the interactors. Continuous user interactions will eventually result in user exhaustion [17], negatively influencing the reliability of the obtained feedback. Therefore, user interactions should not only be considered time consuming but users also need to be treated as a scarce resource – the interaction should be kept to a required minimum. A common way to overcome this problem is to combine interactive optimization algorithms with a surrogate-based approach [22,23]. Surrogate models are typically used as a proxy of functions which are either unknown or extremely time consuming to compute. Classic candidates for such surrogates are machine learning (ML) models. In [16] a survey of popular surrogate functions is provided, ranging from polynomial regression [8] to more sophisticated techniques such as neural networks [12] and support vector regression [7].

Our approach also exhibits similarities to so-called *interactive ML* approaches. Such methods are typically used when the number of training samples is not sufficient to properly train an ML model. To compensate this problem, a human is used as a guide to reduce the search space during the learning phase [13]. One way to reduce the search space is to reduce the number of features considered during the learning phase [5]. Our approach can be considered as an interactive ML algorithm in the sense that the ML models in use are continuously improved and corrected through user evaluations.

4 Cooperative Optimization Algorithm

The proposed solution framework, which we call *Cooperative Optimization Algorithm (COA)*, consists of the following interacting components: an *evaluation component (EC)*, an *optimization component (OC)*, a *feedback component (FC)*, and a *solution management component (SMC)*.

The FC is responsible for selecting or deriving for each user individual candidate solutions that are then presented to him for evaluation. A user $u \in U$ gives feedback to a proposed solution x by stating how much of his demand would actually be satisfied at which locations, i.e., he returns the values $d(u, v, x)$, $v \in V$.

The EC provides a function for evaluating candidate solutions. This in particular also includes intermediate solutions that are not directly evaluated by the users. Should for a solution x the values $d(u, v, x)$ be known for each user $u \in U$ and all locations $\{v \in V \mid x_v = 1\}$, we can calculate the exact objective value $f(x)$ according to (3). Otherwise, we resort to a surrogate objective function $\tilde{f}(x)$ based on an ML model that estimates the real objective value based on the information gained from the users' feedback so far. The respective learning mechanism also is part of the EC.

One call of the OC solves the SPDP by using the EC's current surrogate objective function for evaluating any intermediate solutions and yields one or more optimal or close-to-optimal solutions w.r.t. the current state of the EC. Note that the surrogate objective function never changes during one call of the OC. Instead, the OC is called repeatedly in each major iteration of the framework after having obtained new user feedback and re-trained the EC.

Finally, the SMC efficiently stores and manages information on all candidate solutions that are relevant for more than one of the above components and in particular also the solutions for which users have given feedback.

Figure 1 illustrates the communication between the components, and Algorithm 1 shows how the components of the framework interact with each other within the main program. The algorithm starts with the FC by presenting each user the solution in which all locations are selected, i.e., $x_v = 1$, $v \in V$. The information acquired in this way over all users serves as initial training data for the EC. The following subsections describe each component's functionality in more detail.

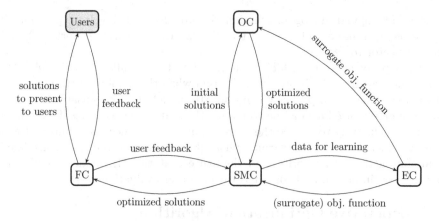

Fig. 1. Components of the framework and their interaction. The framework consists of the feedback component (FC), the evaluation component (EC), the optimization component (OC), and the solution management component (SMC). Users can interact with the framework via the FC.

4.1 Solution Management Component

As the number of locations $|V|$ is usually much larger than the number of service stations that can actually be built in a feasible solution, candidate solutions in the SMC are compactly represented by the subset of locations chosen for setting up service points. Let us denote this set of locations by $s(x) = \{v \in V \mid x_v = 1\}$.

Next to storing all considered candidate solutions, the SMC also maintains for each solution the users for which the exact fulfilled demand is already known, and for each user $u \in U$ a set of so far identified *relevant locations* V_u, which includes any location for which the user has indicated positive demand in at least one solution. Note that the complete set of relevant locations in general is unknown. However, it is the task of the FC to choose the solutions presented to the users in such a way that as many relevant locations as possible are identified (see Sect. 4.2).

Another important task of the SMC is to derive a user's demand for scenarios where this can be efficiently achieved through logical implications from previous scenario evaluations. For example, any scenario that is a superset of a scenario with already maximum fulfillable demand $d_u^{\max}$ will also achieve this maximum.

4.2 Feedback Component

In each major generation of Algorithm 1, the FC generates for each user $u \in U$ an individual set of solutions to evaluate. It is assumed here that any user u evaluates each solution in a completely rational way so that the total fulfilled demand is maximal. The number of user evaluations of solutions needs to be kept as low as possible to avoid user fatigue [17], we cannot ask real users to evaluate hundreds of solutions. Thus, each solution presented to a user must be

---

Algorithm 1. Basic Framework

Input : an instance of the SPDP
Output: a solution $x = (x_v)_{v \in V} \in \{0,1\}^n$
1: $X_{OC} = \{(1, \ldots, 1)\}$ // initial solution, later best solution(s) from the OC **while** *no termination criterion satisfied* **do**
2:      **Feedback Component:**
3:          **for** $u \in U$ **do**
4:              determine set of solutions X'_u to be evaluated by u from X_{OC} and further data in the SMC;
5:              let user u evaluate X'_u, update the SMC with evaluated solutions from X'_u;
6:          **end for**
7:      **Evaluation Component:**
8:          train surrogate objective function $\tilde{f}(x)$ with data from the SMC;
9:          re-evaluate all solutions stored in the SMC with the new surrogate objective function;
10:     **Optimization Component:**
11:         adopt so far best solutions from the SMC as initial solutions;
12:         $X_{OC} \leftarrow$ perform optimization using the EC's surrogate objective function $\tilde{f}(x)$;
13:         when possible, calculate exact $f(x)$ for $x \in X_{OC}$;
14:         store the solution(s) from X_{OC} in the SMC;
15: **end while**
16: **return** *overall best found solution x^* w.r.t. $\tilde{f}$;*

---

non-redundant and meaningful in the way that we likely obtain new knowledge on his needs that is valuable for finding an overall optimal solution.

It appears natural that a solution presented to a user should be similar to the best solutions identified so far by the OC or, otherwise, provide substantial information gain on locations that are potentially interesting for the user. Next to finding new relevant locations for a user, it is also necessary to gain information on the relationship between relevant locations.

We apply the following combination of strategies for compiling a set of at most κ solutions presented to each user $u \in U$, where κ is a strategy parameter and is set to 15 in the experiments performed for this article.[1] Note that solutions generated with these strategies may not necessarily be feasible, which, however, does not immediately matter for the intended purpose.

Best Solution Strategy. Let $f(x)$ be the evaluation function in which we consider exact fulfilled demands $d(u, v, x)$ and let $\tilde{f}$ be the surrogate function, which will be defined in Sect. 4.3. Select the γ_1 best feasible solutions w.r.t. to f and the γ_2 best feasible solutions w.r.t. the surrogate function $\tilde{f}$ (see Sect. 4.3) for which no exact total fulfilled demand is known yet for user u. Hereby, γ_1 and γ_2 are

---

[1] All parameter values stated in the text have been tuned in comprehensive preliminary tests.

strategy parameters, which are both set to 2 in the experiments performed for this article. This strategy clearly focuses on getting exact evaluations for the currently most promising solutions.

Irrelevant Locations Strategy. This strategy focuses purely on finding new relevant locations for a user u, which might lead to good alternative solutions. For this purpose a solution in which the locations in $V \setminus V_u$ are selected is generated.

Best Solution Mutation Strategy. This strategy is a combination of the previous strategies and tries to gain information on the relationship between locations by replacing a subset of $s(x) \cap V_u$ of a solution x obtained from the best solution strategy for a user u with a set of locations for which it is so far unclear if they are relevant: A new solution x' is constructed from a copy of an existing solution x by setting $x_v = 0$ with $v \in s(x) \cap V_u$ with a certain probability ξ for each v, where ξ is a strategy parameter which is set to 0.5 in the experiments throughout this article. Afterwards, we set $x_v = 1$ for n' uniformly at random chosen locations $v \in V \setminus s(x)$ with n' being chosen uniformly at random from $\{0, \ldots, |V| - |s(x)|\}$.

4.3   Evaluation Component

The EC provides the means for evaluating solutions, in particular also temporary solutions generated within the OC. Within the OC the objective value of a solution is estimated by a surrogate objective function $\tilde{f}(x)$, which is defined in accordance to $f(x)$ but makes use of *estimated fulfilled demands*

$$\tilde{d}(u, v, x) = \begin{cases} 0 & \text{if } v \notin V_u \vee x_v = 0 \\ \max(0, g_{u,v}(x)) & \text{else} \end{cases} \tag{4}$$

for each user $u \in U$ and each location $v \in V$, where $g_{u,v}(x)$ represents an ML model trained by all solutions so far evaluated by user u. Note that our definition of $\tilde{d}(u, v, x)$ ensures that conditions (1) are always fulfilled and gives function $g_{u,v}(x)$ more freedom in the sense that it may return negative values, which are mapped to zero, and arbitrary values in case of $x_v = 0$. Furthermore, for any location v for which user u has so far never indicated any positive fulfilled demand in any solution, i.e., for any so far not relevant location $v \in V \setminus V_u$, $g_{u,v}(x) = 0$ is assumed and no ML model needs to be maintained.

Similarly to [21], we use an adaptive surrogate function in the sense that the ML model for each $g_{u,v}(x)$ is initially simple and is upgraded to a higher complexity model during the course of the algorithm when the error of the model – measured in terms of the usual mean squared error MSE of $\tilde{d}(u, v, x)$ – exceeds a certain threshold τ. In this way we stay as efficient as possible from a computational perspective and substantially reduce problems with overfitting.

Our initial choice for $g_{u,v}(x)$ is the linear model (LM)

$$g_{u,v}^{\text{LM}}(x) = w_{u,v} + \sum_{v' \in V_u \setminus \{v\}} w'_{u,v,v'} \cdot x_{v'}. \tag{5}$$

Ridge regression with a penalization factor of one is used for determining the weights $w_{u,v}$ and $w'_{u,v,v'}$. This model is sufficient for covering simple scenarios where users have independent demands that can be fulfilled at specific locations. Furthermore, it can even accurately represent the case where for a user one demand can be fulfilled at a specific primary location or, with a possibly reduced amount, at one alternative location if no service station is set up at the primary location. More complex dependencies, including in particular more than one alternative location, are, however, beyond the capability of the LM.

In this case, which is detected by a remaining MSE of $\tilde{d}(u, v, x)$ larger than a threshold $\tau = 0.075$, we turn to a neural network, starting with a single layer perceptron with a leaky rectified linear unit (ReLU) activation function [18]. This simple neural network realizes the function

$$g_{u,v}^{\text{NN}}(x) = \phi\left(g_{u,v}^{\text{LM}}(x)\right) \text{ with } \phi(S) = \begin{cases} S & \text{if } S \geq 0 \\ \varepsilon \cdot S & \text{else.} \end{cases} \tag{6}$$

The leaky ReLU activation function ϕ serves as an extension of the LM in the sense that this perceptron takes actively into account that satisfied demands cannot be negative. Due to this non-linearity, it can accurately represent scenarios in which for a user a demand can be fulfilled at an arbitrary number of ordered alternative locations, where a service station at one of these locations will only fulfill a certain amount of the demand when no station is set up at any of the preceding alternative locations in the order. We use here the leaky ReLU function with parameter $\varepsilon = 0.01$ which returns small negative values in case the sum S is negative.

While the above perceptron is already more powerful, it is still limited when a user has more than one demand that can be fulfilled partly at the same locations, or more generally, when the different demands are related in some way. Again, we detect the insufficiency of the perceptron by a MSE that exceeds τ and turn in this case to a more complex feed forward neural network with one hidden layer that contains initially two hidden neurons. These neurons again make use of the leaky ReLU activation function, while the single output layer neuron corresponds to a simple summation of the inputs. Initially, we use two hidden neurons and increase this number until, after training, either the MSE does not exceed τ anymore or a maximum of $\lambda = 6$ hidden neurons is reached.

Note that the solutions used for training the models are not required to be feasible, since user evaluations do not consider the budget at all.

4.4 Optimization Component

Remember that the OC is called in each major iteration of the whole framework and makes use of the current surrogate function provided by the EC, which does not change during each individual run of the OC. The OC is thus supposed to return an optimal or close-to-optimal solution w.r.t. the current surrogate function.

The OC is implemented as a Variable Neighborhood Search (VNS) and follows the classical scheme from [20]. It consists of a randomized construction heuristic, a local search part, and a shaking mechanism for escaping local optima. The initial solution is generated via the randomized construction heuristic that considers all locations in random order and sets up a station at a location as long as the budget is not exceeded.

Our local search follows a first improvement strategy and utilizes a two-exchange neighborhood structure, in which a location in the solution is replaced by a location not contained in the solution. The VNS only considers feasible solutions, hence, we skip all moves in the neighborhood resulting in budget constraint violations. Moreover, after each feasible move, we try to additionally improve the solution by adding stations at further locations to the solution in a random order as long as the budget allows it.

Shaking removes stations from a number of uniformly selected random locations and then iteratively adds stations to other locations in a uniform random order, such that the solution stays feasible and no more locations can be added. The number of stations to be removed corresponds to the index of the shaking neighborhood and varies from one to two.

The VNS terminates if no better solution has been found within 40 iterations.

5   Experimental Evaluation

We test the suggested framework in a proof-of-concept manner on artificial benchmark scenarios using an idealized simulation of all user interaction. To a large degree, the proposed framework is independent of the concrete application as long as our general problem formulation is suitable. The machine learning models in the EC, however, were already designed with a few assumptions on user requirements, as they appear, for example, in the context of setting up charging stations for EVs: Users would like to have certain needs associated with use cases fulfilled that are related to particular geographic locations, such as their home and/or work address or other places they visit regularly. While ideally respective service stations would be set up at precisely these locations, the respective demands can to a certain degree also be fulfilled by service stations located in the vicinity. The degree (amount) of fulfilled demand, however is assumed to decrease with the distance. In this way, we implicitly also consider the convenience for the users. It is generally assumed that for fulfilling a demand, a user always uses a station that is closest to the demand's original location.

5.1   Benchmark Scenarios

The primary parameters for our benchmark scenarios are the number of potential locations for service stations n and the number of users m, and we consider here the combinations $n = 50, 60, \ldots, 100$ with $m = 50$ and $n = 50$ with $m = 50, 60, \ldots, 100$. The n locations correspond to points in the Euclidean plane with coordinates chosen uniformly at random from the grid $\{0, \ldots, L-1\}^2$, where

$L = \lceil 10\sqrt{n} \rceil$ is the underlying width and height. The fixed costs c_v as well as the variable costs z_v for setting up a service station at each location $v \in V$ are uniformly chosen at random from $\{50, \ldots, 100\}$. The budget is assumed to be $B = \lceil 7.5 \cdot n \rceil$ so that about 10% of the stations with average costs can be set up. We assume each of the m users $u \in U$ has ρ_u so-called *use cases*, where ρ_u is chosen randomly according to a shifted Poisson distribution with offset one and expected value three. Each of these use cases $i = 1, \ldots, \rho_u$ is associated with a particular geographical location $r_{u,i} \in \{0, \ldots, L-1\}^2$ and a respective demand $d^*_{u,i}$ that could ideally be fulfilled there. This demand can, for example, be the expected number of usages of a service point in a time period. Here, we choose each $d^*_{u,i}$ uniformly at random from $\{5, \ldots, 50\}$. In a real scenario, the locations where demand arises will clearly not be uniformly distributed over the whole considered geographic area. There will be more popular regions as well as less popular ones. We want to consider this aspect and therefore first choose $\alpha = \lceil (L/50)^2 \rceil$ *attraction points* A with uniform random coordinates from $\{0, \ldots, L-1\}^2$ and then derive the location for each use case from a uniformly selected attraction point $(a_x, a_y) \in A$ by

$$r_{u,i} = (\lfloor \mathcal{N}(a_x, 20) \rfloor \bmod L, \lfloor \mathcal{N}(a_y, 20) \rfloor \bmod L), \tag{7}$$

where $\mathcal{N}(\cdot, \cdot)$ denotes a random value sampled from a normal distribution with the respectively given mean value and standard deviation.

For each use case $i = 1, \ldots, \rho_u$ of each user $u \in U$, demand is always only fulfilled at the closest location $v^{\text{clst}}_{u,i}(x) \in V$ w.r.t. the Euclidean distance where a service station is set up in the current candidate solution x (ties are broken according to the locations' natural order) and when a maximum distance, chosen here as 12, is not exceeded. We further assume an exponential decay of the fulfilled demand in dependence of the distance and round down to the closest integer, obtaining

$$d_i(u, v, x) = \begin{cases} \lfloor d^*_{u,i} \cdot e^{-||r_{u,i} - v^{\text{clst}}_{u,i}(x)||/10} \rfloor & \text{if } v = v^{\text{clst}}_{u,i}(x) \wedge ||r_{u,i} - v|| \le 12 \\ 0 & \text{else,} \end{cases} \tag{8}$$

where $|| \cdot ||$ denotes the L^2 norm. These fulfilled demands for each use case i are finally summed up in order to obtain the overall fulfilled demands $d(u, v, x) = \sum_{i=1} d_i(u, v, x)$ for each user $u \in U$ and location $v \in V$ under candidate solution x. Finally, the prize earned for each unit of fulfilled demand in our objective function is assumed to be $p = 50$.

For each combination of n and m 30 independent scenarios were created, and they are available at https://www.ac.tuwien.ac.at/research/problem-instances. The benchmarks were also specifically designed with the ability in mind to calculate proven optimal solutions to which we will compare the solutions of our framework. Exploiting the complete knowledge of the data and specific structure in a "white-box" manner allows the problem to be expressed as mixed-integer linear programming (MIP) model, which we solved with the MIP-solver Gurobi[2].

---

[2] http://www.gurobi.com/.

5.2   Computational Experiments

The OC was implemented in C++, compiled with GNU G++ 5.5.0, while the remaining components of the framework were realized in Python 3.7. For linear regression scikit-learn 0.17 was used and for the perceptrons and neural networks Keras 2.2.2 on top of Theano 1.0.2 (without GPU support). The perceptrons and neural networks were trained with the adam optimizer (learning rate 0.1) over 5000 epochs with a batch size of 32 in order to minimize the MSE. All test runs have been executed on an Intel Xeon E5-2640 v4 with 2.40 GHz machine. Our framework terminated when no improved solution could be found over five iterations or when the CPU-time limit of 7200 s had been reached and returned the overall best found solution x^* w.r.t. the approximate evaluation function $\tilde{f}$.

Table 1 lists average results of COA over all 30 instances for each considered combinations of n and m. Each line shows the average number of iterations n_{it}, the average of the exact objective values of the finally returned solutions $\overline{f(x^*)}$ and the corresponding optimal solutions $\overline{f(x_{opt})}$ obtained from Gurobi by solving the white-box MIP, the average optimality gap %-gap and corresponding standard deviation $\sigma_{\%\text{-gap}}$, where the %-gap is calculated for a final solution x^* of COA in relation to an optimal solution x_{opt} as %-gap $= 100\% \cdot (f(x_{opt}) - f(x^*))/f(x_{opt})$, the average percentage error of the surrogate function values of final solutions $\overline{\%\text{-}\Delta\tilde{f}}$ (serving as an indicator for the quality of the surrogate function) and the corresponding standard deviation $\sigma_{\%\text{-}\Delta\tilde{f}}$, with $\%\text{-}\Delta\tilde{f} = 100\% \cdot |\tilde{f}(x^*) - f(x^*)|/f(x^*)$, and the median of the computation times in seconds. The table shows that COA finds near optimal solutions for almost all instances. Average final gaps to optimal solutions are always less than 0.6%. Moreover, the surrogate function predicts the actual user demand at least for the final solutions excellently; for all instance groups the average percentage error of the surrogate function values is below 0.55%. The percentage errors and computation times tend to slightly increase with an increasing number of users, while the optimization gaps show no such behavior. Neither the number of users, nor the number of locations seems to have an impact on the optimality gaps, indicating that our algorithm is able to also solve larger instances with a similar solution quality. The generally rather high computation times can be explained by the large number of machine learning models that need to be trained in each iteration but also by the fact that the OC is implemented as pure black-box optimization.

Next, in Fig. 2, we take a closer look at the percentage errors of the surrogate function. The boxplots in Fig. 2 show the distribution of $\%\text{-}\Delta\tilde{f}$ for all instances. The figure shows that the percentage error of the final surrogate function of an instance is almost always below 1%. As observed in Table 1, one can see a slight increase of the percentage errors as the number of users increases, i.e. there seems to be a correlation of the size of the percentage errors and the number of users.

Figures 3a–b visualize for an exemplary run with $n = 100$ and $m = 50$ the best solutions w.r.t. $\tilde{f}$ at the first and at the last iteration of COA, respectively. Blue dots show the locations of users' use cases with their sizes indicating the respective maximal satisfiable demands $d^*_{u,i}$. Diamonds show the potential

Table 1. Average results of COA.

n	m	$\overline{n_{it}}$	$f(x^*)$	$f(x_{opt})$	%-gap	$\sigma_{\text{%-gap}}$	%-$\Delta\tilde{f}$	$\sigma_{\text{%-}\Delta\tilde{f}}$	time[s]
50	50	10	56499	56717	0.34	0.67	0.19	0.14	2063
50	60	10	67134	67467	0.38	1.05	0.31	0.19	2594
50	70	11	78562	78845	0.21	0.58	0.33	0.34	2936
50	80	10	88003	88283	0.28	0.51	0.35	0.32	3522
50	90	10	98408	98961	0.56	0.89	0.37	0.34	3867
50	100	11	106604	107020	0.28	0.48	0.41	0.44	4424
60	50	11	59189	59354	0.46	1.43	0.53	1.35	2335
70	50	11	61685	62097	0.49	0.93	0.38	0.48	2447
80	50	12	64425	64690	0.49	0.73	0.27	0.28	2904
90	50	11	66689	66870	0.27	0.40	0.23	0.19	2946
100	50	13	71030	71229	0.26	0.44	0.28	0.27	3889

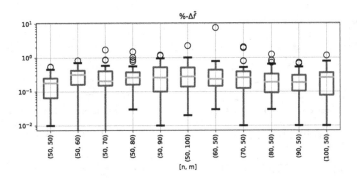

Fig. 2. Distributions of the percentage errors of the surrogate function values of final solutions.

locations of service stations V, with the larger ones with the discs corresponding to those chosen in the best solution of the iteration. The actually fulfilled demand of a service station is indicated by the size of the diamonds, and the discs illustrate the covered area in respect to the maximum distance of 12. We can see that already the solution obtained in the first iteration is quite meaningful. Although the final solution is similar at the first glance, a closer look reveals a significantly better coverage of demands in the final solution.

This observation is also confirmed by Fig. 3c showing the corresponding development of the best solution's exact objective value over the iterations in comparison to the optimal solution value $f(x_{opt})$. We can see that already the solution of the first iteration has a relatively high objective value, which is continuously improved in few iterations until the optimum is almost reached.

Finally, Fig. 3d shows the distribution of the model sizes of the surrogate function's underlying machine learning models at the final iteration of COA. A model size of zero refers to LMs, size one to perceptrons, and larger sizes to neural networks with the respective number of neurons in the hidden layer. The distribution shown in Fig. 3d is typical for all instances tested. It shows that the majority of machine learning models is made up of LMs and perceptrons. Larger size neural networks are rarely needed. However, the figure also shows a small peak at the largest neural network with six neurons in its hidden layer. This peak is caused by unpopular service point locations resulting in training data in which most customer demands are zero. The neural networks often fail to properly learn such data, however, on the other hand, as these locations are the least popular service point locations, they usually have no large impact on the final solution.

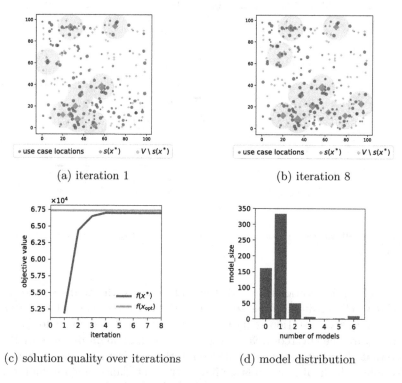

(a) iteration 1   (b) iteration 8

(c) solution quality over iterations   (d) model distribution

Fig. 3. An exemplary run with $n = 100$ and $m = 50$: (a–c) best solutions at different iterations and (d) exact objective value of best solution per iteration.

6 Conclusion

We proposed a cooperative algorithm framework for distributing service points within a geographical area in mobility applications under incomplete information. Instead of estimating user demands by combining a variety of more or less reliable sources, our method directly incorporates potential customers in the optimization process. Our proof-of-concept implementation is still based on comparatively simple components. Nevertheless we could show that the machine learning models in our evaluation component are able to learn the non-trivial user behavior of all our benchmark scenarios reliably after relatively few user interactions, and the optimization is able to indeed find solutions with only small remaining optimality gaps. The careful derivation of the candidate solutions to be presented to the users in the feedback component also plays a particularly important role.

In future work we will investigate the approach on more complex scenarios such as bike sharing systems, where a use case always relates to two, usually different locations for renting and returning a bike, respectively. Considering capacity limits and different possible configuration options for the service points is another practically highly relevant aspect. Another challenge is to improve the scalability of the approach towards more potential locations and more users. To this end it seems necessary to replace the individual machine learning models we currently have for each user and each location by a more integrated approach. Even though, we use a surrogate function to unburden the customers from evaluating too many solutions, the current number of solutions a user needs to evaluate is still very high. Additional efforts need to be made to further reduce this number. Last but not least, improvements should also be possible in the optimization component.

References

1. Cavadas, J., Homem, G.d.A.C., Gouveia, J.: A MIP model for locating slow-charging stations for electric vehicles in urban areas accounting for driver tours. Transp. Res. Part E Log. Transp. Rev. **75**, 188–201 (2015)
2. Chen, T., Kockelman, K.M., Khan, M.: The electric vehicle charging station location problem: a parking-based assignment method for Seattle. In: 92nd Annual Meeting of the Transportation Research Board in Washington DC (2013)
3. Church, R., ReVelle, C.: The maximal covering location problem. Pap. Reg. Sci. **32**, 101–118 (1974)
4. Cornuéjols, G., Nemhauser, G.L., Wolsey, L.A.: The uncapacitated facility location problem. In: Mirchandani, P.B., Francis, R.L. (eds.) Discrete Location Theory, pp. 119–171. Wiley, New York (1990)
5. Fails, J.A., Olsen Jr., D.R.: Interactive machine learning. In: Proceedings of the 8th International Conference on Intelligent User Interfaces, pp. 39–45. IUI 2003. ACM, New York (2003)
6. Farahani, R.Z., Hekmatfar, M.: Facility Location: Concepts, Models, Algorithms and Case Studies. Springer, Heidelberg (2009). https://doi.org/10.1007/978-3-7908-2151-2

7. Forrester, A., Andras, S., Keane, A.: Engineering Design via Surrogate Modelling: A Practical Guide. Wiley, New York (2008)

8. Forrester, A.I., Keane, A.J.: Recent advances in surrogate-based optimization. Prog. Aerosp. Sci. **45**(1), 50–79 (2009)

9. Frade, I., Ribeiro, A., Gonçalves, G., Antunes, A.: Optimal location of charging stations for electric vehicles in a neighborhood in Lisbon, Portugal. Transp. Res. Rec. J. Transp. Res. Board **2252**, 91–98 (2011)

10. Gavalas, D., Konstantopoulos, C., Pantziou, G.: Design and management of vehicle-sharing systems: a survey of algorithmic approaches. In: Obaidat, M.S., Nicopolitidi, P. (eds.) Smart Cities and Homes, pp. 261–289. Elsevier (2016)

11. Laporte, G., Nickel, S., da Gama, F.S. (eds.): Location Science. Springer, Cham (2015). https://doi.org/10.1007/978-3-319-13111-5

12. Haykin, S.: Neural Networks: A Comprehensive Foundation. Prentice Hall, Englewood Cliffs (1998)

13. Holzinger, A.: Interactive machine learning for health informatics: when do we need the human-in-the-loop? Brain Inf. **3**(2), 119–131 (2016)

14. Kameda, H., Mukai, N.: Optimization of charging station placement by using taxi probe data for on-demand electrical bus system. In: König, A., Dengel, A., Hinkelmann, K., Kise, K., Howlett, R.J., Jain, L.C. (eds.) KES 2011. LNCS (LNAI), vol. 6883, pp. 606–615. Springer, Heidelberg (2011). https://doi.org/10.1007/978-3-642-23854-3_64

15. Kim, H.S., Cho, S.B.: Application of interactive genetic algorithm to fashion design. Eng. Appl. Artif. Intell. **13**(6), 635–644 (2000)

16. Koziel, S., Ciaurri, D.E., Leifsson, L.: Surrogate-based methods. In: Koziel, S., Yang, X.S. (eds.) Computational Optimization, Methods and Algorithms. Studies in Computational Intelligence, vol. 356, pp. 33–59. Springer, Heidelberg (2011). https://doi.org/10.1007/978-3-642-20859-1_3

17. Llorà, X., Sastry, K., Goldberg, D.E., Gupta, A., Lakshmi, L.: Combating user fatigue in iGAs: partial ordering, support vector machines, and synthetic fitness. In: Proceedings of the 7th Annual Conference on Genetic and Evolutionary Computation, GECCO 2005, pp. 1363–1370. ACM, New York (2005)

18. Maas, A.L., Hannun, A.Y., Ng, A.Y.: Rectifier nonlinearities improve neural network acoustic models. In: Workshop on Deep Learning for Audio, Speech and Language Processing, ICML 2013 (2013)

19. Meignan, D., Knust, S., Frayret, J.M., Pesant, G., Gaud, N.: A review and taxonomy of interactive optimization methods in operations research. ACM Trans. Interact. Intell. Syst. **5**(3), 17:1–17:43 (2015)

20. Mladenović, N., Hansen, P.: Variable neighborhood search. Comput. Oper. Res. **24**(11), 1097–1100 (1997)

21. Shi, L., Rasheed, K.: ASAGA: an adaptive surrogate-assisted genetic algorithm. In: Proceedings of the 10th Annual Conference on Genetic and Evolutionary Computation, GECCO 2008, pp. 1049–1056. ACM, New York (2008)

22. Sun, X.Y., Gong, D., Li, S.: Classification and regression-based surrogate model-assisted interactive genetic algorithm with individual's fuzzy fitness. In: Proceedings of the 11th Annual Conference on Genetic and Evolutionary Computation, GECCO 2009, pp. 907–914. ACM, New York (2009)

23. Sun, X., Gong, D., Jin, Y., Chen, S.: A new surrogate-assisted interactive genetic algorithm with weighted semisupervised learning. IEEE Trans. Cybern. **43**(2), 685–698 (2013)

A Binary Algebraic Differential Evolution for the MultiDimensional Two-Way Number Partitioning Problem

Valentino Santucci[1(✉)], Marco Baioletti[2], Gabriele Di Bari[2],
and Alfredo Milani[2]

[1] Department of Humanities and Social Sciences,
University for Foreigners of Perugia, Perugia, Italy
`valentino.santucci@unistrapg.it`
[2] Department of Mathematics and Computer Science, University of Perugia,
Perugia, Italy
{`marco.baioletti,alfredo.milani`}`@unipg.it, gabriele.dibari@unifi.it`

Abstract. This paper introduces MADEB, a Memetic Algebraic Differential Evolution algorithm for the Binary search space. MADEB has been applied to the Multidimensional Two-Way Number Partitioning Problem (MDTWNPP) and its main components are the binary differential mutation operator and a variable neighborhood descent procedure. The binary differential mutation is a concrete application of the abstract algebraic framework for the binary search space. The variable neighborhood descent is a local search procedure specifically designed for MDTWNPP. Experiments have been held on a widely accepted benchmark suite and MADEB is experimentally compared with respect to the current state-of-the-art algorithms for MDTWNPP. The experimental results clearly show that MADEB is the new state-of-the-art algorithm in the problem here investigated.

Keywords: Binary algebraic differential evolution ·
Multidimensional Two-Way Number Partitioning Problem ·
Variable neighborhood descent

1 Introduction

In this paper we propose a memetic binary variant of the Algebraic Differential Evolution (ADE) algorithm [1], namely, MADEB, for the multidimensional two-way number partitioning problem (MDTWNPP) [2].

MDTWNPP has been originally introduced in [2] as a multidimensional generalization of the more famous number partitioning problem (NPP) [3].

An instance of MDTWNPP is given as a set S of n real-valued vectors of dimension d, i.e., $S = \{w_k \in \mathbb{R}^d : 1 \leq k \leq n\}$. The objective is to partition S into two subsets S_0 and S_1, i.e., $S_0 \cup S_1 = S$, $S_0 \cap S_1 = \emptyset$, and such that the

© Springer Nature Switzerland AG 2019
A. Liefooghe and L. Paquete (Eds.): EvoCOP 2019, LNCS 11452, pp. 17–32, 2019.
https://doi.org/10.1007/978-3-030-16711-0_2

sum of the vectors in S_0 is as close as possible to that of S_1. Formally, given the partition $\{S_0, S_1\}$, the objective function to be minimized is

$$f(S_0, S_1) = \max_{1 \leq j \leq d} \left| \sum_{w_i \in S_0} w_{ij} - \sum_{w_i \in S_1} w_{ij} \right|, \tag{1}$$

where w_{ij} is the j-th component of the vector w_i.

Clearly, any partition $\{S_0, S_1\}$ can be represented as a n-length bit-string x, i.e., $x \in \mathbb{B}^n$, in such a way that $x(i) = 0$ when $w_i \in S_0$, and $x(i) = 1$ when $w_i \in S_1$. Therefore, MDTWNPP is a binary optimization problem.

MDTWNPP is NP-hard because, for $d = 1$, it reduces to the optimization variant of classical NPP. Moreover, as described in [2], MDTWNPP is even more difficult than NPP, because the computational complexity of its instances does not decrease together with the ratio between the number of bits needed to express any input element and the quantity of input elements (as it has been observed to happen in NPP instances [3,4]). This greater complexity has been also experimentally confirmed in [5], where it has been observed that CPLEX, applied to an integer linear programming model for MDTWNPP, has never been able to compute a better lower bound than the trivial one, i.e., zero.

For these reasons, meta-heuristic algorithms have recently been applied to MDTWNPP [5–7]. Following this line of research, here we introduce a binary variant of the Algebraic Differential Evolution (ADE) algorithm specifically designed for MDTWNPP. ADE has been originally introduced in [1,8] and, though it has been extensively applied to permutation-based optimization problems [9–13], the abstract algebraic framework at its bases works with any finitely generated group [1,14].

In this paper, we show how it is possible to see the binary search space $\mathbb{B}^n$ as a finitely generated group, thus we formally derive an algebraically and geometrically consistent implementation of the differential mutation operator for the $\mathbb{B}^n$ space. It is worthwhile to note that our proposal is substantially different from the other binary DE schemes in the literature. For example, [15] and [16] propose two binary DEs that mainly work in the continuous space and decode back the numerical vectors to bit-strings as soon as they are needed.

Besides, we introduce two new algorithmic components: (i) a modified initialization scheme, (ii) an effective variable neighborhood descent scheme for MDTWNPP.

Due to these considerations, we will call our algorithm MADEB (Memetic Algebraic Differential Evolution for the Binary space) in the rest of paper, that has been organized as follows. Section 2 provides a brief review of the state-of-the-art algorithms for MDTWNPP. The main scheme of MADEB is introduced in Sect. 3. Its key operators, i.e., the binary algebraic differential mutation and the variable neighborhood descent, are introduced and motivated in Sects. 4 and 5. Computational experiments have been described and analyzed in Sect. 6. Finally, conclusion are drawn in Sect. 7 which also depicts some possible future lines of research.

2   Related Work

In this section we describe the main approaches used to solve MDTWNPP. Although this problem is a direct generalization of the number partitioning problem [3], many techniques used to solve the latter cannot be extended to the multidimensional case, like, for instance, the Karmarkar-Karp heuristic (see [2]).

The MDTWNPP has been defined in [2], where an integer linear programming formulation is provided. Here, a set of 210 benchmark instances have been randomly generated and solved by using the linear programming solver CPLEX.

A memetic algorithm (MA) has been defined in [6] for solving a generalization of the MDTWNPP, called multidimensional multiway number partitioning problem, in which the vectors must be partitioned in $p \geq 2$ subsets. The MA is a genetic algorithm which produces, at each generation, 10 offsprings for every population individual by means of crossover, mutation and local search operators. The best offsprings are then selected for the next-generation population. The local search method uses a k-change neighborhood, for $k = 1, 2, 3$, where k denotes the number of bits changed by a single move. The authors performed computational experiments also in the case of $p = 2$, i.e., in the MDTWNPP problem. The results show that MA outperforms CPLEX in most instances.

Two other meta-heuristic approaches have been introduced in [7].

The first one is a VNS-like procedure which operates on a incumbent solution x, represented as a bit-string. A series of increasing neighborhoods $N_k(x)$ are employed in the shaking phase, along with a local search whose elementary step is to flip a 0-bit and a 1-bit of x. This corresponds to swap two vectors: one vector moves from partition S_0 to partition S_1, while the other one moves in the opposite direction. The generic neighborhood $N_k(x)$ is defined as the set of all the bit-strings having Hamming distance k from x. The parameter k is increased from $k_{min} = 2$ to $k_{max} = \min\{30, \lfloor n/4 \rfloor\}$ (in a circular way) at every iteration where the new solution does not improve the incumbent one.

The second meta-heuristic uses an Electromagnetism-like (EM) approach. Solutions are represented as real vectors with components in $[0, 1]$. The partition associated to a solution is defined by applying a threshold equal to 0.5. At each generation, every solution undergoes to a local search and scale operators, then all the solutions are moved according to "electromagnetic forces" that can be attractive or repulsive and depend on the objective values.

The experimental results show that VNS and EM obtained comparable performances and both outperform MA and CPLEX.

Finally, in [5], a GRASP equipped with an Exterior Path Relinking method is described. The algorithm evolves a set of solutions, called elite set. At each step, the GRASP procedure produces a new solution by means of two operations: construction and local improvement. The former operation builds-up a solution by means of a greedy method, while the latter iteratively improves the incumbent solution by using a possibly restricted local search in the space of the 2-change neighborhood. Then, the Path Relinking phase explores a path from the new solution s_i to a randomly selected solution s_G in the elite set (Interior PR) or beyond s_G (Exterior PR), returning the best solution found in the path.

The configuration with the Exterior Path Relinking, i.e., GRASP+ePR, resulted in better performances and, compared with VNS and CPLEX, it almost always outperformed them, thus representing the current state-of-the-art algorithm for MDTWNPP.

3   The General Scheme of MADEB

The classical Differential Evolution (DE) [17,18] is a popular and effective evolutionary algorithm for continuous optimization problems that iteratively evolves a population of numerical vectors by means of three genetic operators: differential mutation, crossover, and selection. In particular, the differential mutation is widely considered to be the key component of DE [19]. Despite its proven effectiveness in numerical optimization [20,21], most of the DE schemes for combinatorial problems are not so effective, perhaps, because their search behavior is loosely connected to the working mechanisms of continuous DE (see, for example, [15,16]). In order to fill this gap, in [1], an original algebraic framework has been introduced in order to design a differential mutation for combinatorial search spaces in such a way that it consistently simulates the behavior of its numerical counterpart. The framework abstractly defines the design of the combinatorial differential mutation in any finitely generated group. In the previous works, implementations for the permutations search space have been proposed [8,12]. In MADEB, we introduce the first instantiation of the abstract combinatorial differential mutation for binary search spaces.

Therefore, MADEB evolves a population of N bit-strings by iterative applications of the following operators: binary algebraic differential mutation, variable neighborhood descent, and selection. Its general scheme is depicted in Algorithm 1.

Algorithm 1. General scheme of MADEB

1: **function** MADEB($N, init$)
2:     Initialize N bit-strings $\{x_1, \ldots, x_N\}$ by means of the $init$ procedure
3:     **while** termination condition has not been satisfied **do**
4:         **for** $i = 1$ to N **do**
5:             $y_i \leftarrow AlgebraicDifferentialMutation(x_i)$
6:             $z_i \leftarrow VariableNeighborhoodDescent(y_i)$
7:         **end for**
8:         **for** $i = 1$ to N **do**
9:             $x_i \leftarrow Selection(x_i, z_i)$
10:        **end for**
11:        **if** x_{best} has not been updated in the last $1\,000$ generations **then**
12:            Reinitialize the bit-strings in $\{x_1, \ldots, x_N\} \setminus \{x_{best}\}$ by using $init$
13:        **end if**
14:    **end while**
15:    **return** the best visited bit-string x_{best}
16: **end function**

Two population initialization procedures are considered. When $init = R$, the bit-strings are randomly initialized as usual, i.e., every bit $x_i(j)$, with $1 \leq i \leq N$ and $1 \leq j \leq n$, is initialized to 0 or 1, with probability 0.5. When $init = U$, a probability value $p_i \in [0, 1]$ is randomly generated for every individual x_i, and its bits $x_i(j)$ are independently set to 1 with probability p_i, or 0 otherwise. In this way, the expected number of 1-bits throughout the population individuals is uniformly distributed. The rationale behind the $init = U$ scheme is to allow a more diverse population initialization.

For every population individual x_i, *AlgebraicDifferentialMutation* generates a mutant y_i as follows:

$$y_i \leftarrow x_i \oplus F_i \odot (x_{r_1} \ominus x_{r_2}), \tag{2}$$

where: $F_i > 0$ is the DE scale factor, $r_1, r_2 \in [1, N]$ are randomly generated indexes different between them and with respect to i, and the $\oplus, \ominus, \odot$ are the binary versions of our algebraic operators that are introduced and discussed in Sect. 4. It is worthwhile to note that the scale factor is self-adapted during the evolution by means of the popular jDE scheme [22]. In particular: every individual has its own F_i value; the differential mutation is computed using a scale factor randomly generated in $[0.1, 2]$ with probability 0.1, or F_i otherwise; if the trial individual replaces x_i during the selection, then F_i is updated with the employed scale factor.

After the differential mutation, the mutant y_i undergoes a local search phase. The local search procedure *VariableNeighborhoodDescent* adopts two neighborhood definitions and generates the trial individual z_i in such a way that z_i is a local minimum of both neighborhoods. *VariableNeighborhoodDescent* is further described in Sect. 5.

The *Selection* procedure replaces x_i with z_i if and only if $f(z_i) < f(x_i)$, where f is the fitness function defined in Eq. (1). Moreover, if the best population individual x_{best} has not been updated during the last 1 000 generations, the population, except x_{best}, is randomly reinitialized by means of the chosen *init* procedure.

Summarizing, MADEB requires only two parameters to be set: the population size N, and the initialization procedure $init \in \{R, U\}$.

4   Algebraic Differential Mutation for the Binary Space

The binary algebraic differential mutation is the main component of MADEB. In the following, after briefly recalling the abstract algebraic framework (originally introduced in [1,12]), we introduce its implementation for the binary space. This allows to present the binary algebraic differential mutation, that is further analyzed in the last part of this section.

4.1   Abstract Algebraic Framework

A combinatorial search space is a set X of discrete solutions. X forms a group if there exists a binary operation $\star$ between the elements of X such that: $\star$ is associative, there exists a unique neutral element, and any $x \in X$ has a unique inverse $x^{-1} \in X$. The group is finitely generated if there exists a subset $H \subseteq X$ such that every $x \in X$ can be decomposed as $x = h_1 \star h_2 \star \cdots \star h_l$, for some $h_1, h_2, \ldots, h_l \in H$. H is the *generating set* of X and its elements are called *generators*.

A search space representable by a finitely generated group $(X, \star, H)$ can be visualized by its Cayley graph $\mathcal{C}(X, \star, H)$, whose vertices are the solutions in X, and there is an arc from $x \in X$ to $y \in X$, labeled by $h \in H$, if and only if $x \star h = y$.

Interestingly, the labels in a shortest path from x to y corresponds to the sequence of generators in a minimal decomposition of $x^{-1} \star y$. We denote the length of the minimal decomposition (or, equivalently, of the shortest path) by $|y^{-1} \star x|$.

These aspects allow us to define the algebraic operators $\oplus, \ominus, \odot$ that simulate, in the Cayley graph (i.e., the combinatorial search space), the geometric behavior of their numerical counterparts.

We define $x \oplus y := x \star y$ and $y \ominus x := x^{-1} \star y$. The sum $s = x \oplus y$ can be interpreted in the Cayley graph as starting from vertex x and moving (towards s) by iteratively choosing the arcs corresponding to a minimal decomposition of y. The difference $d = y \ominus x$ can be interpreted as the composition of the generators in a shortest path going from x to y, thus d is a synthetic representation of this path. As in the numerical case, $x \oplus (y \ominus x) = x \star (x^{-1} \star y) = y$.

Moreover, given a scalar $F \geq 0$, the multiplication $F \odot x$ can be described as follows. Let us interpret x as the representation of a shortest path between two arbitrary vertices of the Cayley graph, then, when $F \leq 1$ (or $F > 1$), $F \odot x$ represents, a truncation (or an extension) of the path represented by x. Algebraically, this geometric interpretation can be encoded by requiring $z = F \odot x$ to satisfy these properties: (i) $|z| = \lceil F \cdot |x| \rceil$, (ii) if $F \leq 1$, the sequence of generators in a minimal decomposition of z is a prefix of the sequence of generators in a minimal decomposition of x, or (iii) vice versa, when $F > 1$. Since, in general, minimal decompositions (thus shortest paths) are not unique, there can be multiple $z \in X$ satisfying these properties, therefore, we define $\odot$ as a stochastic operator that randomly returns an element with the given properties.

Summarizing, the geometric interpretation of the algebraic differential mutation of Eq. (2) is as follows. Let first compute the generators in a (shortest) path from x_{r_2} to x_{r_1}; truncate (or extend) the sequence of generators considering the scalar F; start from the vertex x_i and move, towards the result y_i, by following one by one the arcs labeled with the generators in the truncated (or extended) sequence.

4.2 Binary Algebraic Differential Mutation

$\mathbb{B}^n$ is the set of all the bit-strings of length n and it forms an Abelian group with respect to the bit-wise XOR operation $\veebar$. Indeed, $\veebar$ is commutative and associative, the zero bit-string $\mathbf{0}$ is the neutral element, and the inverse of each $x \in \mathbb{B}^n$ is itself, i.e., $x = x^{-1}$.

Given $x \in \mathbb{B}^n$, we denote by $x(i)$ the i–th bit of x, for $1 \leq i \leq n$. We recall that the Hamming weight $|x|$ is the number of 1-bits of x. The Hamming distance between two bit-strings x and y is the number of positions i such that $x(i) \neq y(i)$.

Moreover, $\mathbb{B}^n$ is a finitely generated group. Its generating set U is composed by the n bit-strings with Hamming weight equal to 1, thus the generic generator $u_i \in U$, for $1 \leq i \leq n$, is such that $u_i(i) = 1$, while the rest of its bits are 0.

Therefore, any $x \in \mathbb{B}^n$ can be written as $x = u_{i_1} \veebar u_{i_2} \veebar \ldots \veebar u_{i_l}$, where $i_1, i_2, \ldots, i_l$ are the indexes of the 1-bits of x. Clearly, $l = |x|$. The decomposition is minimal and unique, up to reordering the indexes $i_1, i_2, \ldots, i_l$. We exploit this property and we represent the minimal decomposition of $x \in \mathbb{B}^n$ as the set $U_x = \{u_i \in U : x(i) = 1\}$. Note anyway that any ordering of the generators in U_x is a sequence that fulfills the abstract framework definitions.

Importantly, note that, for each $x \in \mathbb{B}^n$, the application of the generator u_i to x, i.e., $x \veebar u_i$, corresponds to flipping the i–th bit of x.

As any other finitely generated group, $(\mathbb{B}^n, \veebar, U)$ has its associated Cayley graph. Since $\veebar$ is commutative and each bit-string is the inverse of itself, the Cayley graph $\mathcal{C}(\mathbb{B}^n, \veebar, U)$ reduces to an indirected labelled graph, which corresponds to the hypercube with n vertices, where all the pairs of bit-strings, differing in a single bit i, are connected by an edge labelled with u_i.

Therefore, by following the abstract definitions given in Sect. 4.1, it is now possible to concretely derive the operations $\oplus, \ominus, \odot$ for the binary space.

The addition $\oplus$ is defined as $x \oplus y := x \veebar y$ for $x, y \in \mathbb{B}^n$. The subtraction uses the property that $x^{-1} = x$ for each $x \in \mathbb{B}^n$, hence $y \ominus x := x \veebar y$. Note that, in this particular group, $\oplus$ and $\ominus$ coincide.

Given $F \geq 0$ and $x \in \mathbb{B}^n$, the stochastic scalar multiplication $F \odot x$ is defined as randomly selecting a $z \in \mathbb{B}^n$ such that its decomposition U_z: (i) has size $k = \lceil F \cdot |x| \rceil$, and (ii) when $F \leq 1$, $U_z \subseteq U_x$, while, (iii) if $F > 1$, $U_z \supseteq U_x$. It is easy to see that any ordering of the generators in U_z satisfies the three properties described in Sect. 4.1.

When $F \leq 1$, z is computed by randomly selecting one of the $\binom{|x|}{k}$ subsets of size k of U_x.

When $F > 1$, U_z is computed as $U_x \cup A$, where A is randomly selected among the $\binom{n-|x|}{k-|x|}$ subsets of $U \setminus U_x$. This is equivalent to randomly flipping some of the 0-bits of x.[1]

Since this is a concrete realization of the abstract framework of Sect. 4.1, all the geometric considerations previously provided are valid also in this search space. In particular, it is interesting to note that the elementary search moves here considered are bit-flip moves, that is, the most common and natural moves adopted in the binary search space.

---

[1] For this reason, $|F \odot x|$ cannot be larger than n, thus we truncate F to $F_{\max}^{(x)} = \frac{n}{|x|}$ whenever $F > F_{\max}^{(x)}$.

4.3   Search Characteristics of the Binary Differential Mutation

Here we analyze the binary implementation of the algebraic differential mutation provided in Eq. (2).

We start by describing the computation of the mutant $y_i \leftarrow x_i \oplus F \odot (x_{r_1} \ominus x_{r_2})$ according to the definitions introduced in Sect. 4.2. Lets first compute the bit-wise XOR between x_{r_1} and x_{r_2}. This corresponds to selecting the positions where the bits of x_{r_1} and x_{r_2} differ. A subset (or a superset) of these positions is computed by considering the scalar F. Finally, x_i is moved, towards the result y_i, by flipping all the bits of x_i at the positions previously computed.

As an illustrative example, let consider $x_i = (1010)$, $x_{r_1} = (1001)$, $x_{r_2} = (1110)$, and $F = 0.66$. The difference between x_{r_1} and x_{r_2} is computed as $d = x_{r_1} \ominus x_{r_2} = x_{r_1} \veebar x_{r_2} = (0111)$. Its Hamming weight is $|d| = 3$, thus we have to randomly select $\lceil F \cdot |d| \rceil = 2$ 1-bits from d. Let's choose the last two 1-bits of d, thus $F \odot d = (0011)$. Finally, $y_i = x_i \oplus (F \odot d) = x_i \veebar (F \odot d) = (1001)$.

It is interesting to note that, by denoting the Hamming distance with d_H, we have that $d_H(x_i, y_i) = F \cdot d_H(x_{r_1}, x_{r_2})$, i.e., the amount of perturbation applied to x_i is decided by the scale factor F and the two randomly selected population individuals x_{r_1} and x_{r_2}. Moreover, even the positions where the bit-flips are applied are computed by means of x_{r_1} and x_{r_2}. This behavior is quite analogous to what happens when classical differential mutation is applied in the continuous space [17].

However, the structural characteristics of the binary space can introduce some issues in the search behavior of the binary differential mutation. Indeed, binary $\oplus$ and $\ominus$ are actually the bit-wise XOR operator. Moreover, the bit-strings x_i, x_{r_1}, x_{r_2}, involved in the differential mutation formula, are three individuals taken from the current MADEB population. This implies that, when the population reaches consensus on a bit (i.e., when all individuals have their i-th bit set to the same value), it is impossible to flip that bit in the future generations by only using the binary differential mutation with $F \leq 1$. This is the reason of why we use an interval with a right bound larger than 1 for the scale factor F, i.e., $F \in [0.1, 2]$.

Furthermore, it is worthwhile to note that most of the binary crossovers in the literature are somehow special cases of our binary differential mutation. Indeed, a binary crossover between two generic $x, y \in \mathbb{B}^n$ usually computes an offspring $z \in \mathbb{B}^n$ such that the bit value $z(j)$ is equal to $x(j)$ or $y(j)$, for $1 \leq j \leq n$. Possible examples are the uniform crossover, the one-point crossover or the more general k-points crossover [23]. It is easy to see that the computation of such an offspring can be reproduced in the algebraic framework as $z = x \oplus F \odot (y \ominus x)$, by using $F \in [0, 1]$. For instance, the uniform crossover which takes $z(j) = x(j)$ with probability 0.5, or $z(j) = y(j)$ otherwise, for $j = 1, \ldots, n$, is equivalent on average to $z = x \oplus 0.5 \odot (y \ominus x)$. Other crossover operators can be obtained by simply considering different selection strategies (other than the random one) in the $\odot$ definition. Therefore, since a binary crossover is a special case of our binary differential mutation, we have decided to not employ the crossover operator in MADEB.

5   Variable Neighborhood Descent for MDTWNPP

In MADEB, every mutant y undergoes a local search procedure implemented as a variable neighborhood descent scheme.

The *VariableNeighborhoodDescent* procedure uses two different neighborhoods N_1 and N_2. $N_1(y)$ is the classical bit-flip neighborhood, while $N_2(y)$ is the 2-change neighborhood that contains all the bit-strings which can be obtained from y by flipping one 0-bit and one 1-bit.

The two neighborhoods are explored alternatively until no better solution is found as depicted in Algorithm 2.

A first-improvement style is used in every iteration of the neighborhood local searches (lines 4–8 and 9–13 of the pseudocode), i.e., as soon as an improving

---

Algorithm 2. Pseudocode of *VariableNeighborhoodDescent*

1: **function** VARIABLENEIGHBORHOODDESCENT(y)
2:      $imp1 \leftarrow true$
3:      **while** $imp1$ **do**
4:          $z \leftarrow y$
5:          /* Local Search in N_1 */
6:          $imp2 \leftarrow true$
7:          **while** $imp2$ **do**
8:              Randomly permute the solutions in $N_1(y)$
9:              $y' \leftarrow$ the first solution in $N_1(y)$ s.t. $f(y') < f(y)$, or *nil* if no improvement
10:             **if** y' is not *nil* **then**
11:                 $y \leftarrow y'$
12:             **else**
13:                 $imp2 \leftarrow false$
14:             **end if**
15:         **end while**
16:         /* Local Search in N_2 */
17:         $imp2 \leftarrow true$
18:         **while** $imp2$ **do**
19:             Randomly permute the solutions in $N_2(y)$
20:             $y' \leftarrow$ the first solution in $N_2(y)$ s.t. $f(y') < f(y)$, or *nil* if no improvement
21:             **if** y' is not *nil* **then**
22:                 $y \leftarrow y'$
23:             **else**
24:                 $imp2 \leftarrow false$
25:             **end if**
26:         **end while**
27:         /* Main loop operations */
28:         **if** $f(y) = f(z)$ **then**
29:             $imp1 \leftarrow false$
30:         **end if**
31:     **end while**
32:     **return** y
33: **end function**

---

neighbor is found, it is returned. Moreover, the neighbors in N_1 and N_2 are randomly scanned. Finally, it is important to note that the computation of a neighbor solution is not made by scratch, but incrementally with respect to the incumbent solution. Indeed, by keeping track of the two partition sums, it is possible to compute the objective value of a neighbor in both N_1 and N_2 in $\Theta(d)$ time.

It is easy to prove that the overall computational cost for each iteration of the local search in N_1 is $\Theta(n \cdot d)$, while for N_2 is $\Theta(n^2 \cdot d)$.

6   Experiments

The behavior of MADEB in the MDTWNP problem has been experimentally analyzed by considering the benchmark set of 210 instances provided in [2], where the author proposed 5 instances for each problem configuration n, d with $n \in \{50, 100, 200, 300, 400, 500\}$ and $d \in \{2, 3, 4, 5, 10, 15, 20\}$.

Following the methodology adopted in [5], we have divided the instances in two sets: tuning and test instances. The tuning set consists of the first two instances in every problem configuration n, d and it has been adopted in order to select a good setting for MADEB. The rest of the instances have been devoted to experimentally compare the tuned MADEB with the state-of-the-art algorithms for MDTWNPP.

In order to make a fair comparison, as in [5], for each problem instance, MADEB has been executed 10 times with a computational budget of 600 seconds per execution. The experiments were carried out on a machine equipped with an Intel Xeon X5650 processor clocking at 2.67 GHz, which has a very similar computational power with respect to the machine adopted in [5].

In the following, we present the tuning of the MADEB setting and the experimental comparison with the state-of-the-art results for MDTWNPP.

6.1   Experimental Tuning of MADEB

MADEP has two parameters to be set: the population size N, and the initialization procedure $init$. Eight MADEB settings have been analyzed, with $N \in \{50, 100, 150, 200\}$ and $init \in \{R, U\}$, by performing a full factorial analysis on 84 tuning instances.

The average objective values obtained by every MADEB setting, on its executions on every tuning instance, have been recorded. In Table 1, we provide the ranks of the MADEB settings averaged over all the tuning instances.

The best setting is $(N = 50, init = U)$, which reached the lowest average rank of 4.04. Therefore, this is the setting used for the experimental comparison discussed in Sect. 6.2.

Moreover, as recommended in [24], we have conducted a non-parametric Friedman test in order to analyze the statistical differences among the settings. Since the Friedman test shows that there is no significant difference among the eight settings, MADEB can be considered a robust algorithm.

Table 1. Average ranks of MADEB settings in the experimental tuning.

MADEB setting	Avg rank
$N = 50, \text{init} = U$	4.04
$N = 50, \text{init} = R$	4.27
$N = 150, \text{init} = U$	4.40
$N = 100, \text{init} = U$	4.43
$N = 200, \text{init} = R$	4.43
$N = 100, \text{init} = R$	4.45
$N = 200, \text{init} = U$	4.56
$N = 150, \text{init} = R$	4.95

6.2   Comparison with State-of-the-Art MDTWNPP Algorithms

In this section we compare MADEB, using the setting ($N = 50$, $init = U$), with the state-of-the-art MDTWNPP algorithms described in Sect. 2.

As in [5], MADEB has been executed 10 times on every one of the 126 test instances. Its results have been compared with those obtained by GRASP+ePR [5], VNS [7], and CPLEX [2]. The results for the competitor algorithms have been directly obtained from the supplementary material of [5] available online at https://sci2s.ugr.es/MDTWNP.

The performance of each algorithm A, on every instance i, is measured by the commonly adopted average relative percentage deviation (ARPD) [25]:

$$ARPD_i^A = \frac{1}{k} \sum_{j=1}^{k} \frac{(A_i^j - Best_i)}{Best_i} \times 100, \tag{3}$$

where A_i^j is the objective value obtained by the algorithm A in its j-th run on the instance i, and $Best_i$ is the best objective value achieved among all executions of all the algorithms on the problem instance i.

Moreover, in order to detect the statistical differences among the ARPD results, as suggested in [24], the non-parametric Friedman test and the Finner post-hoc procedure have been applied.

For each problem instance, the best objective value and the ARPDs of each algorithm are provided in Table 2. The best ARPD on each instance is reported in bold, while the best objective value is in bold when it has been reached by (at least) an execution of MADEB. The table is divided in six groups according to the different values of n. For any group of instances, the average rank of every algorithm is provided together with a symbol indicating the result of the statistical comparison with MADEB: "+" means that MADEB significantly outperforms the competing algorithm. Moreover, the average ranks aggregated for every value of n and d are shown, respectively, in Figs. 1a and b.

Table 2. Experimental comparison of MADEB with state-of-the-art algorithms.

Problem Instance	Best Obj. Val.	MADEB	GRASP +ePR	VNS	CPLEX
50_2c	**3.083**	45.20	76.86	141.71	1784492.93
50_2d	**1.349**	113.36	46.30	306.27	3115763.60
50_2e	**2.938**	97.80	119.91	189.55	1135570.52
50_3c	**142.935**	1.07	2.78	2.78	232764.59
50_3d	**96.410**	5.07	8.55	4.13	358105.58
50_3e	**211.797**	1.59	1.59	2.45	325728.98
50_4c	**452.007**	0.88	0.81	1.42	398.81
50_4d	**1018.573**	0.30	0.42	0.42	192.14
50_4e	**1202.624**	0.30	0.43	0.51	438.27
50_5c	**1398.771**	0.39	32.16	6.84	334.37
50_5d	**2269.686**	0.16	0.17	0.47	63.77
50_5e	**4360.613**	1.70	4.35	2.83	40.99
50_10c	**14125.537**	7.76	23.15	26.68	71.35
50_10d	**14915.871**	0.06	3.38	23.47	50.64
50_10e	**15356.800**	1.86	7.75	8.64	96.02
50_15c	**29456.850**	4.76	18.34	24.96	78.96
50_15d	**21655.890**	0.00	28.88	59.13	133.79
50_15e	**31800.690**	1.17	5.86	14.54	47.81
50_20c	**50560.860**	0.18	9.32	6.64	34.47
50_20d	**53955.960**	1.90	4.62	6.46	33.10
50_20e	**48281.499**	0.00	21.09	9.86	47.95
Average Rank		1.17	2.17	2.67	4.00
Stat. Comp. vs MADEB			+	+	+
100_2c	**0.768**	307.97	499.79	1129.65	2030759.38
100_2d	**0.001**	61140.00	213560.00	643980.00	3833499900.00
100_2e	**1.067**	14.55	28.88	140.17	4919581.35
100_3c	**231.512**	7.45	15.14	7.58	330.18
100_3d	**243.989**	1.20	2.82	3.49	209597.57
100_3e	**135.305**	4.51	4.51	28.95	867653.59
100_4c	**896.038**	0.65	0.99	1.60	96.57
100_4d	**1090.117**	0.67	0.59	1.57	44.96
100_4e	**508.289**	1.32	1.92	8.74	174.41
100_5c	**2769.736**	1.93	2.21	1.04	63.89
100_5d	**2969.833**	0.17	0.16	0.45	101.24
100_5e	**2448.321**	52.88	54.22	30.54	142.58
100_10c	**11984.020**	15.69	23.46	18.81	145.19
100_10d	**14904.240**	4.35	17.14	12.71	59.72
100_10e	**12182.070**	18.01	32.75	12.60	150.43
100_15c	**31112.332**	3.23	7.74	12.73	37.64
100_15d	**30690.906**	0.47	8.09	13.80	64.74
100_15e	**30250.569**	1.35	6.17	8.41	91.59
100_20c	**46021.607**	7.91	25.64	16.24	60.14
100_20d	**43786.661**	7.53	21.48	16.78	53.39
100_20e	**43923.294**	10.32	23.09	17.57	56.30
Average Rank		1.26	2.31	2.43	4.00
Stat. Comp. vs MADEB			+	+	+
200_2c	**0.690**	91.43	241.54	436.93	6394972.46
200_2d	**0.237**	499.83	1353.50	1846.75	11773317.72
200_2e	**0.451**	138.14	371.84	829.18	6777948.78
200_3c	**1.526**	159.95	209.89	12237.23	36695640.50
200_3d	**77.565**	46.30	52.20	50.38	774080.36
200_3e	**105.284**	55.99	91.57	54.65	199011.93
200_4c	**3.047**	65.25	50.26	22036.19	65850.77
200_4d	**770.115**	30.52	38.66	28.90	255.22
200_4e	**508.023**	73.13	142.78	43.84	313.03
200_5c	**3.052**	79.97	114.95	21439.79	163770.90
200_5d	**2130.675**	11.31	20.91	14.09	160.83
200_5e	**1231.049**	201.39	207.33	100.74	277.90
200_10c	**12785.776**	2.74	23.75	34.85	150.40
200_10d	**17390.299**	3.05	7.60	11.92	43.43
200_10e	**17794.136**	0.03	0.05	2.60	63.35
200_15c	**30467.014**	0.05	0.30	7.79	100.80
200_15d	**22837.984**	16.25	42.28	47.55	118.27
200_15e	**28602.541**	3.10	17.88	19.96	100.96
200_20c	**41990.608**	9.37	19.81	25.52	83.76
200_20d	**41177.738**	4.52	20.56	26.18	83.11
200_20e	**39411.579**	11.22	21.00	34.81	84.52
Average Rank		1.24	2.24	2.52	4.00
Stat. Comp. vs MADEB			+	+	+

Problem Instance	Best Obj. Val.	MADEB	GRASP +ePR	VNS	CPLEX
300_2c	**0.759**	198.30	339.24	657.69	1529412.52
300_2d	**0.433**	272.93	892.89	1378.71	863410.39
300_2e	**0.917**	33.82	244.25	928.60	588013.41
300_3c	**214.729**	4.69	15.82	16.79	109847.42
300_3d	**91.207**	26.42	34.62	51.23	269626.01
300_3e	**167.205**	10.95	41.30	22.88	229009.18
300_4c	**885.348**	1.24	1.52	12.57	127.67
300_4d	**765.805**	30.47	41.12	27.66	126.28
300_4e	**736.131**	22.18	23.63	22.56	135125.25
300_5c	**901.749**	1.11	1.32	126.17	430.23
300_5d	**2115.419**	0.52	19.29	7.01	178.64
300_5e	**1543.701**	0.32	0.57	25.15	177.60
300_10c	**15971.771**	0.06	0.13	5.23	68.21
300_10d	**15812.711**	7.84	17.66	10.35	59.91
300_10e	**15291.777**	2.14	10.39	12.64	142.85
300_15c	**22220.712**	27.14	42.70	42.66	138.53
300_15d	**28891.210**	13.59	18.29	20.76	77.21
300_15e	**28754.344**	1.65	4.61	20.00	81.33
300_20c	**34247.615**	18.34	41.50	48.49	77.57
300_20d	**42351.439**	8.14	20.75	21.77	67.56
300_20e	**37132.464**	19.65	32.14	34.36	108.58
Average Rank		1.05	2.29	2.67	4.00
Stat. Comp. vs MADEB			+	+	+
400_2c	**0.884**	123.39	167.83	694.05	1419357.01
400_2d	**1.164**	87.21	297.39	715.05	2524.57
400_2e	**0.406**	340.64	636.31	1786.97	3609013.30
400_3c	**66.576**	153.85	256.07	139.45	434203.65
400_3d	**212.073**	4.11	8.28	31.62	135968.24
400_3e	**155.001**	10.63	30.76	24.73	245907.45
400_4c	**807.262**	10.43	11.19	25.30	218.84
400_4d	**806.920**	1.25	11.33	11.85	123275.06
400_4e	**725.451**	23.47	28.36	26.62	116.37
400_5c	**902.131**	1.27	1.55	129.45	342.50
400_5d	**1667.575**	49.72	49.88	21.07	239.24
400_5e	**1595.032**	0.77	1.15	26.64	239.76
400_10c	**14503.050**	7.97	15.32	17.85	113.16
400_10d	**13723.390**	5.42	10.77	16.53	75.20
400_10e	**11571.600**	39.64	40.86	35.61	183.04
400_15c	**26815.632**	3.60	12.16	21.41	108.86
400_15d	**26395.830**	16.89	25.34	23.54	99.42
400_15e	**30105.307**	1.85	2.98	12.46	78.08
400_20c	**45374.378**	2.07	11.29	10.40	63.91
400_20d	**41384.465**	4.87	15.65	21.38	59.66
400_20e	**39541.150**	9.05	21.45	25.04	76.03
Average Rank		1.14	2.33	2.52	4.00
Stat. Comp. vs MADEB			+	+	+
500_2c	**0.578**	64.71	372.73	728.53	2333117.99
500_2d	**0.202**	603.86	941.58	3457.08	3.96
500_2e	**0.205**	611.37	1070.78	1679.32	7324290.24
500_3c	**96.343**	39.49	95.09	55.76	391556.89
500_3d	**1.522**	330.93	421.16	4294.56	20003973.59
500_3e	**3.056**	84.79	144.71	3267.62	9173885.60
500_4c	**633.460**	47.44	69.00	85.70	103.02
500_4d	**3.057**	159.54	194.40	23031.45	72864.34
500_4e	**552.412**	3.76	53.09	44.45	153.71
500_5c	**1891.494**	23.11	51.26	43.38	168.56
500_5d	**3.053**	159.97	154.91	42282.41	999801.74
500_5e	**2526.657**	0.39	6.41	7.77	176.26
500_10c	**14567.870**	19.45	21.82	16.04	60.34
500_10d	**12875.690**	0.12	2.69	13.25	86.87
500_10e	**15527.300**	4.07	11.73	9.50	64.39
500_15c	**27109.630**	0.08	6.49	23.23	55.07
500_15d	**29550.430**	0.31	7.97	10.06	85.27
500_15e	**25010.130**	20.61	25.06	30.11	106.69
500_20c	**37587.384**	10.89	25.91	32.81	87.31
500_20d	**43249.879**	5.36	10.22	15.77	49.51
500_20e	**39011.410**	11.10	21.95	22.43	96.33
Average Rank		1.14	2.24	2.76	3.86
Stat. Comp. vs MADEB			+	+	+

The results provided in Table 2 clearly show that MADEB outperforms all the other algorithms. In particular, our algorithm reached the best objective value in 76 instances over 126. Moreover, MADEB obtained the best ARDP value on 106 instances, i.e., about the 84% of the benchmark suite. Importantly,

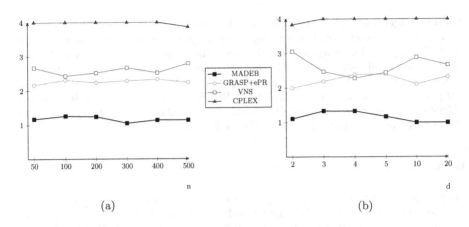

Fig. 1. Average ranks aggregated by the values of (a) n and (b) d

the statistical test has detected noticeable differences between our algorithm and the competitors: in every group of instances, MADEB is significantly better than all the other algorithms. Indeed, the worst Finner p-value (obtained in the comparison with GRASP+ePR on the group of instance $n = 200$) is about $2 \cdot 10^{-6}$, which is well below the commonly considered critical threshold 0.05 [24].

Finally, the better performances of MADEP with respect to its competitors are apparent also from the data provided in the two graphs of Figs. 1a and b. Clearly, MADEB always obtained the best average rank, both when the instances are grouped by the values of n or the values of d.

7   Conclusions and Future Work

In this paper we have described a Memetic Algebraic Differential Evolution algorithm for solving Binary combinatorial optimization problems (MADEB). The main components of MADEB are: a binary algebraic differential mutation, and a variable neighborhood descent procedure. In particular, the binary differential mutation has been introduced as an instantiation of the abstract algebraic framework [1,26–28].

MADEB has been applied to the MDTWNP problem. Experiments have been held on widely adopted benchmark instances, and the experimental results show that our approach has outperformed the state-of-the-art MDTWNPP algorithms.

As a future lines of research we are considering to apply MADEB to other binary optimization problems, like the knapsack problem and its variations, and to develop an algebraic particle swarm scheme [29] for the binary space.

Another possible future work is to extend MADEB for solving the multi-way generalization of MDTWNPP, as defined in [6], which would however require a major change to MADEB, because the solutions are no more binary strings.

Acknowledgement. The research described in this work has been partially supported by: the research grant "Fondi per i progetti di ricerca scientifica di Ateneo 2019" of the University for Foreigners of Perugia under the project "Algoritmi evolutivi per problemi di ottimizzazione e modelli di apprendimento automatico con applicazioni al Natural Language Processing"; and by RCB-2015 Project "Algoritmi Randomizzati per l'Ottimizzazione e la Navigazione di Reti Semantiche" and RCB-2015 Project "Algoritmi evolutivi per problemi di ottimizzazione combinatorica" of Department of Mathematics and Computer Science of University of Perugia.

References

1. Santucci, V., Baioletti, M., Milani, A.: Algebraic differential evolution algorithm for the permutation flowshop scheduling problem with total flowtime criterion. IEEE Trans. Evol. Comput. **20**(5), 682–694 (2016). https://doi.org/10.1109/TEVC.2015.2507785
2. Kojić, J.: Integer linear programming model for multidimensional two-waynumber partitioning problem. Comput. Math. Appl. **60**(8), 2302–2308 (2010). http://www.sciencedirect.com/science/article/pii/S0898122110005882
3. Mertens, S.: The easiest hard problem: number partitioning. Comput. Complex. Stat. Phys. **125**(2), 125–139 (2006)
4. Corus, D., Oliveto, P.S., Yazdani, D.: Artificial immune systems can find arbitrarily good approximations for the NP-hard partition problem. In: Proceedings of 15th International Conference on Parallel Problem Solving from Nature-PPSN XV - Part II, pp. 16–28 (2018)
5. Rodriguez, F.J., Glover, F., García-Martínez, C., Martí, R., Lozano, M.: Grasp with exterior path-relinking and restricted local search for the multidimensional two-way number partitioning problem. Comput. Oper. Res. **78**, 243–254 (2017). http://www.sciencedirect.com/science/article/pii/S0305054816302209
6. Pop, P.C., Matei, O.: A memetic algorithm approach for solving the multidimensional multi-way number partitioning problem. Appl. Math. Model. **37**(22), 9191–9202 (2013). http://www.sciencedirect.com/science/article/pii/S0307904X13002692
7. Kratica, J., Kojić, J., Savić, A.: Two metaheuristic approaches for solving multidimensional two-way number partitioning problem. Comput. Oper. Res. **46**, 59–68 (2014). http://www.sciencedirect.com/science/article/pii/S0305054814000045
8. Santucci, V., Baioletti, M., Milani, A.: A differential evolution algorithm for the permutation flowshop scheduling problem with total flow time criterion. In: Bartz-Beielstein, T., Branke, J., Filipič, B., Smith, J. (eds.) PPSN 2014. LNCS, vol. 8672, pp. 161–170. Springer, Cham (2014). https://doi.org/10.1007/978-3-319-10762-2_16
9. Santucci, V., Baioletti, M., Milani, A.: Solving permutation flowshop scheduling problems with a discrete differential evolution algorithm. AI Commun. **29**(2), 269–286 (2016). https://doi.org/10.3233/AIC-150695
10. Santucci, V., Baioletti, M., Milani, A.: An algebraic differential evolution for the linear ordering problem. In: Companion Material Proceedings of Genetic and Evolutionary Computation Conference, GECCO 2015, pp. 1479–1480 (2015). https://doi.org/10.1145/2739482.2764693

11. Baioletti, M., Milani, A., Santucci, V.: Linear ordering optimization with a combinatorial differential evolution. In: Proceedings of 2015 IEEE International Conference on Systems, Man, and Cybernetics, SMC 2015, pp. 2135–2140 (2015). https://doi.org/10.1109/SMC.2015.373

12. Baioletti, M., Milani, A., Santucci, V.: An extension of algebraic differential evolution for the linear ordering problem with cumulative costs. In: Handl, J., Hart, E., Lewis, P.R., López-Ibáñez, M., Ochoa, G., Paechter, B. (eds.) PPSN 2016. LNCS, vol. 9921, pp. 123–133. Springer, Cham (2016). https://doi.org/10.1007/978-3-319-45823-6_12

13. Baioletti, M., Milani, A., Santucci, V.: MOEA/DEP: an algebraic decomposition-based evolutionary algorithm for the multiobjective permutation flowshop scheduling problem. In: Liefooghe, A., López-Ibáñez, M. (eds.) EvoCOP 2018. LNCS, vol. 10782, pp. 132–145. Springer, Cham (2018). https://doi.org/10.1007/978-3-319-77449-7_9

14. Baioletti, M., Milani, A., Santucci, V.: Learning Bayesian networks with algebraic differential evolution. In: Auger, A., Fonseca, C.M., Lourenço, N., Machado, P., Paquete, L., Whitley, D. (eds.) PPSN 2018. LNCS, vol. 11102, pp. 436–448. Springer, Cham (2018). https://doi.org/10.1007/978-3-319-99259-4_35

15. Wang, L., Fu, X., Mao, Y., Menhas, M.I., Fei, M.: A novel modified binary differential evolution algorithm and its applications. Neurocomputing **98**, 55–75 (2012). http://www.sciencedirect.com/science/article/pii/S0925231212004316

16. Pampara, G., Engelbrecht, A.P., Franken, N.: Binary differential evolution. In: 2006 IEEE International Conference on Evolutionary Computation, pp. 1873–1879, July 2006

17. Storn, R., Price, K.: Differential evolution-a simple and efficient heuristic for global optimization over continuous spaces. J. Global Optim. **11**(4), 341–359 (1997). https://doi.org/10.1023/A:1008202821328

18. Milani, A., Santucci, V.: Asynchronous differential evolution. In: 2010 IEEE Congress on Evolutionary Computation (CEC 2010), pp. 1–7 (2010). https://doi.org/10.1109/CEC.2010.5586107

19. Price, K., Storn, R.M., Lampinen, J.A.: Differential Evolution: A Practical Approach to Global Optimization. Springer, Heidelberg (2006). https://doi.org/10.1007/3-540-31306-0

20. Das, S., Suganthan, P.N.: Differential evolution: a survey of the state-of-the-art. IEEE Trans. Evol. Comput. **15**(1), 4–31 (2011)

21. Das, S., Mullick, S.S., Suganthan, P.: Recent advances in differential evolution-an updated survey. Swarm Evol. Comput. **27**, 1–30 (2016). http://www.sciencedirect.com/science/article/pii/S2210650216000146

22. Brest, J., Greiner, S., Boskovic, B., Mernik, M., Zumer, V.: Self-adapting control parameters in differential evolution: a comparative study on numerical benchmark problems. IEEE Trans. Evol. Comput. **10**(6), 646–657 (2006)

23. Pavai, G., Geetha, T.V.: A survey on crossover operators. ACM Comput. Surv. **49**(4), 72:1-72:43 (2016). http://doi.acm.org/10.1145/3009966

24. Derrac, J., García, S., Molina, D., Herrera, F.: A practical tutorial on the use of nonparametric statistical tests as a methodology for comparing evolutionary and swarm intelligence algorithms. Swarm Evol. Comput. **1**(1), 3–18 (2011). http://www.sciencedirect.com/science/article/pii/S2210650211000034

25. Ceberio, J., Irurozki, E., Mendiburu, A., Lozano, J.A.: A distance-based ranking model estimation of distribution algorithm for the flowshop scheduling problem. IEEE Trans. Evol. Comput. **18**(2), 286–300 (2014)

26. Baioletti, M., Milani, A., Santucci, V.: Algebraic particle swarm optimization for the permutations search space. In: Proceedings of 2017 IEEE Congress on Evolutionary Computation (CEC 2017), pp. 1587–1594 (2017). https://doi.org/10.1109/CEC.2017.7969492

27. Baioletti, M., Milani, A., Santucci, V.: Automatic algebraic evolutionary algorithms. In: Pelillo, M., Poli, I., Roli, A., Serra, R., Slanzi, D., Villani, M. (eds.) WIVACE 2017. CCIS, vol. 830, pp. 271–283. Springer, Cham (2018). https://doi.org/10.1007/978-3-319-78658-2_20

28. Baioletti, M., Milani, A., Santucci, V.: Algebraic crossover operators for permutations. In: 2018 IEEE Congress on Evolutionary Computation (CEC 2018), pp. 1–8 (2018). https://doi.org/10.1109/CEC.2018.8477867

29. Santucci, V., Milani, A.: Particle swarm optimization in the EDAs framework. In: Gaspar-Cunha, A., Takahashi, R., Schaefer, G., Costa, L. (eds.) Soft Computing in Industrial Applications. Advances in Intelligent and Soft Computing, vol. 96, pp. 87–96. Springer, Heidelberg (2011). https://doi.org/10.1007/978-3-642-20505-7_7

A New Representation in Genetic Programming for Evolving Dispatching Rules for Dynamic Flexible Job Shop Scheduling

Fangfang Zhang$^{(\boxtimes)}$, Yi Mei, and Mengjie Zhang

School of Engineering and Computer Science,
Victoria University of Wellington, P.O. Box 600, Wellington 6140, New Zealand
{fangfang.zhang,yi.mei,mengjie.zhang}@ecs.vuw.ac.nz

Abstract. Dynamic flexible job shop scheduling (DFJSS) is a very important problem with a wide range of real-world applications such as cloud computing and manufacturing. In DFJSS, it is critical to make two kinds of real-time decisions (i.e. the routing decision that assigns machine to each job and the sequencing decision that prioritises the jobs in a machine's queue) effectively in the dynamic environment with unpredicted events such as new job arrivals and machine breakdowns. Dispatching rule is an ideal technique for this purpose. In DFJSS, one has to design a routing rule and a sequencing rule for making the two kinds of decisions. Manually designing these rules is time consuming and requires human expertise which is not always available. Genetic programming (GP) has been applied to automatically evolve more effective rules than the manually designed ones. In GP for DFJSS, different features in the terminal set have different contributions to the decision making. However, the current GP approaches cannot perfectly find proper combinations between the features in accordance with their contributions. In this paper, we propose a new representation for GP that better considers the different contributions of different features and combines them in a sophisticated way, thus to evolve more effective rules. The results show that the proposed GP approach can achieve significantly better performance than the baseline GP in a range of job shop scenarios.

Keywords: Representation · Dispatching rules
Dynamic flexible job shop scheduling · Genetic programming

1 Introduction

Job shop scheduling (JSS), as an important optimisation problem, has received a great deal of attention from both academics and industry researchers. It captures practical and challenging issues in real world scheduling tasks such as managing grid/cloud computing [1] and designing manufacturing processes [2]. JSS aims

© Springer Nature Switzerland AG 2019
A. Liefooghe and L. Paquete (Eds.): EvoCOP 2019, LNCS 11452, pp. 33–49, 2019.
https://doi.org/10.1007/978-3-030-16711-0_3

to make a schedule to process a number of jobs with a set of machines. Each job consists of a sequence of operations which need to be processed one by one. Classical JSS assumes that one operation can be processed on only one specific machine. Thus, the task is to schedule the operations in the queue of the machines.

Flexible job shop scheduling (FJSS) is different from the classic JSS in that each operation can be processed by multiple candidate machines. Therefore, FJSS includes two sub-tasks, which are machine assignment and operation sequencing. Machine assignment is to select an appropriate machine for each operation from its candidate machines. Operation sequencing is to determine the order of processing the allocated jobs in each machine to obtain feasible and satisfactory solutions. FJSS is NP-hard [3].

In practice, the JSS problems are typically in *dynamic* environment. For instance, the jobs arrive over time and their attributes are not completely known until they arrive in the shop floor. There are also other types of dynamic events in JSS problem such as order cancellations [4] and machine breakdowns [5]. In this paper, we focus on dynamic new job arrivals because it is the most frequent and common factor in the shop floor. The challenge of dynamic flexible job shop scheduling (DFJSS) is how to capture both the machine assignment (*routing*) decision and operation sequencing (*sequencing*) decision simultaneously along with the new jobs arriving over time.

Exact approaches to search for optimal solutions, such as dynamic programming [6] and branch-and-bound [7], are too time-consuming and inapplicable for solving large scale JSS problems. Therefore, heuristic search methods such as tabu search [8] and genetic algorithm [9] have been commonly adopted to find "near-optimal" solutions in a reasonable time. However, heuristic search methods are not suitable for solving DFJSS problems because of their lack of ability to react to the dynamic events in real time. Dispatching rules are promising in this case because of their low time complexity, the ease of implementation and the ability to cope with dynamic situations in the job shop. Since the term *dispatching rule* has been used in different contexts, it is worth highlighting that the concept of a dispatching rule in DFJSS consists of a routing rule and a sequencing rule (i.e. two kinds of rules). The dispatching rules are normally designed manually, which is very time-consuming and requires human expertise which is not always available. In addition, many manually designed dispatching rules are relatively simple and normally restricted to some specific assumptions [10] and have difficulties in handling complex practical scenarios [11,12]. Genetic programming (GP) has been proven to be a dominating method to automatically design dispatching rules for JSS [11–14]. In order to evolve both the routing rule and sequencing rule for DFJSS, GP was hybridised with the cooperative co-evolution framework to co-evolve the two kinds of rules [15,16].

In GP for DFJSS, different features in the terminal set have different contributions to the decision making. For example, the feature named WIQ (work in queue, i.e. the total processing time of operations in the queue of a machine) is known to be a dominating feature for making the routing decision, as intuitively a machine with lighter workload should be preferred [17]. On the other hand, machine ready time is another important feature for routing decisions, i.e., it tends to be better to assign the operation to a machine that can become idle in the earliest time. However, the contribution of machine ready time should be smaller than WIQ. Intuitively, a machine with a lighter workload but a later ready time should still be better than a machine with a heavier workload but an earlier ready time. However, the existing GP approaches fail to properly combine the features in accordance with their contributions. As a result, the evolved routing rules usually focus too much on WIQ, but overlook the other non-dominating features. The schedules obtained may become ineffective over the longer term when facing the real-world shop environments.

In this paper, we aim to develop a new representation for routing rule to help GP evolve more effective rules for DFJSS. In particular, we consider the following research objectives:

- Develop a novel component that can appropriately take the information of workload of machines into consideration.
- Propose a novel representation that tends to take into account the effect of the dominating feature WIQ and other non-dominating features properly.
- Verify the effectiveness of proposed GP approach with new representation by comparing its performance with the baseline GP.
- Analyse the rules evolved by the proposed GP approach.

2   Background

2.1   Dynamic Flexible Job Shop Scheduling

Given a set of machines $M = \{M_1, M_2, ..., M_m\}$ and jobs $J = \{J_1, J_2, ..., J_n\}$, FJSS aims to determine which machine to process a particular job and which job will be chosen to process next by a particular machine. To be specific, each job J_j has a sequence of l_j ($l_j <= m$) operations $O_j = (O_{j1}, O_{j2}, ..., O_{jl_j})$. Each operation O_{ij} can only be processed by one of its own optional machines $\pi(O_{ij})$ and its processing time $\delta(O_{ij})$ depends on the machine that processes it. Then FJSS is to find an effective schedule subject to the following constraints:

(1) The $(j + 1)th$ operation of J_i (denotes by $O_{i(j+1)}$) can only be processed after its preceding operation O_{ij} has been processed.
(2) Each operation O_{ij} can be processed on one of the corresponding set of machines $\pi(O_{ij}) \subseteq M$ with $\delta(O_{ij})$.
(3) Each machine can process at most one operation at a time.
(4) The scheduling is non-preemptive, i.e. the processing of an operation cannot be stopped or paused until it is completed.

For the dynamic job shop scheduling problem, jobs arrive in the job shop over time and their information can only be known when they arrive.

2.2 Dispatching Rules in Dynamic Flexible Job Shop Scheduling

In DFJSS, a *routing decision situation* will be generated when a new job arrivals or an operation is finished and its subsequent operation becomes a ready operation. A *sequencing decision situation* will be derived when a machine becomes idle and its queue is not empty. Two kinds of dispatching rules are needed in DFJSS, which are routing rule and sequencing rule, respectively. The quality of a schedule depends highly on how well the routing rule and the sequencing rule work together. The routing rule will be triggered to decide which machine to allocate the operation when a routing decision situation is derived. The sequencing rule will be triggered to determine which operation in its queue will be chosen to process next when a sequencing decision situation is derived. Once the trigger conditions are met, the corresponding decisions will be made immediately.

The machine or operation with the highest priority assigned by routing or sequencing rule is identified respectively. Once one operation is finished, its information related to objectives will be recorded to its corresponding job. After all the operations are processed, the recorded information related to all jobs is obtained. Finally, the fitness can be calculated based on the information according to different objectives.

2.3 Related Work

In recent years, GP has been widely used to automatically design dispatching rules for solving JSS problems [18,19]. Tree-based GP is commonly used in many studies [20,21]. In 2007, Tay and Ho [2] proposed a GP approach to evolve priority rules for FJSS with multiple objectives by combining them into a single function. Hildebrandt et al. [11] then used GP to evolve dispatching rules in different simulations for the single objective of meanflow time. The results show that the evolved rules perform very well in different scenarios. However, the early studies only aim to evolve sequencing rule by fixing the routing rule. Cooperation co-evolution was applied to GP to evolve the rules at the same time in [15]. The results show that the evolved rules are more effective.

However, in the standard GP, all the features are considered equally in the terminal set. The WIQ feature is a dominating factor [17,22] that is much more important than other features for routing decisions, therefore, the routing rules tend to select the machine with minimal WIQ. In this case, WIQ tends to be dominant and overweights the other non-dominating features. However, there are many other features in the job shop and they might be less important than WIQ, but still contribute to the routing decisions. Using them improperly could lead to suboptimal performance.

3 The Proposed GP Approach

In order to identify effective combinations of the features with different importance, this paper considers to separate WIQ from other features, thus to expect GP can focus more on non-dominating features during the evolutionary process.

A new representation for routing rule is designed to learn dispatching rules which can take more information of non-dominating features. It is noted that the representation design only applies on routing rule. In this section, the proposed representation is presented first, followed by the designed components.

3.1   Representation

The routing rule is defined as the product of two parts, which are named as component 1 and component 2 (i.e. the details will be given in Sect. 3.2). Component 1 is *predefined* and component 2 is *evolved* by GP approach automatically. The motivation of this design is to separate the WIQ with non-dominating features. It is worth mentioning that a machine with smallest priority value (i.e. highest priority) will be chosen in this paper. Multiplication is used here to combine these two components together as it is a more appropriate combination operator than addition and subtraction. Specifically, the value of component 2 might be much larger than the value of component 1. An example of the routing rule is shown in Fig. 1. The terminals and functions in Fig. 1 will be described in Sect. 4.2.

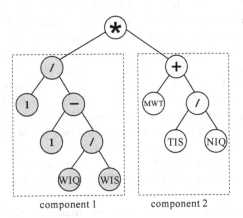

Fig. 1. An example of routing rule with new representation.

3.2   Components Design

The goal of component 1 is to extract the information of the current state solely related to the workload of machines. To this end, component 1 is designed not only to consider the information of WIQ, but also to consider the workload distribution of machines in the system. The workload ratio (i.e. workload of one machine over the entire workload in the system) that can help measure the distribution of allocated jobs of machines. This is to prevent a solution from assigning too much work on a single machine. Thus, the design of component

1 needs to meet two requirements. Firstly, it can be used to indicate the information of workload of individual machines properly. Secondly, it should express the difference of the workload of different machines in the system appropriately.

Under the above consideration, the expression of the proposed component 1 is shown in Eq. (1), where WIQ stands for the total processing time of operations in the queue of a machine, and WIS (work in system. i.e. the total processing time of operations in the queue of all machines) denotes the sum of the workload of all the machines in the job shop. The $component_2$ is the value (V_{GP}) obtained by GP method.

$$component_1 = 1/(1 - WIQ/WIS) \tag{1}$$

$$component_2 = V_{GP} \tag{2}$$

$$Priority = component_1 * component_2 \tag{3}$$

As shown in Eq. (1), component 1 has the following two characteristics. Firstly, it can ensure that all the values obtained by Eq. (1) are larger than 1. It means that component 1 will enlarge the value obtained by the GP approach. In other words, component 1 plays a role of a penalty coefficient. Secondly, the penalty of each machine is different. The heavier the workload of a machine, the severer the punishment is given to that machine. This way, a machine with a lighter workload tends to have a smaller priority value, and is more likely to be selected.

An example of the combination mechanism is shown as follows. We assume there are two candidate machines for one operation. The value obtained by original GP approach are V_{GP1} and V_{GP2}. The workloads of these two machines are denoted by WIQ_1 and WIQ_2. At one decision point, the workload in the entire shop floor (WIS, i.e. work in system) is the same for all the machines. According to the Eqs. (1), (2) and (3), the different situations of the proposed representation are shown in Table 1, where $C()$ stands for the relationship between the corresponding elements. In Table 1, " $=, >, <$ " are used to show the relationship between V_{GP1} and V_{GP2}, WIQ_1 and WIQ_2, $component_1$ and $component_2$, $Priority_1$ and $Priority_2$.

Table 1 shows how the difference between machines workload (WIQ) can influence the final priority value $(Priority)$ under different situations categorised by V_{GP}. All the cases of $V_{GP1} = V_{GP2}$ and most cases of $V_{GP1} > V_{GP2}$ and $V_{GP1} < V_{GP2}$ have the same trend that the WIQ is positively correlated with $component_1$ and $Priority$. This means a machine with lighter workload will get a smaller component 1 and final priority value, and vice versa. This ensures that the priority values obtained by GP is well considered. There are some special cases as follows. When $V_{GP1} > V_{GP2}$ $(V_{GP1} < V_{GP2})$, if $WIQ_1 < WIQ_2$ $(WIQ_1 > WIQ_2)$, the final priority value $(Priority)$ is not sure and depends on the accurate result according to Eqs. (1), (2) and (3). Thus, on one hand, the machine with heavier workload but small V_{GP} still has chance to have a small priority value, and thus be selected. On the other hand, the machine with a poor V_{GP} can also be selected if its workload is very light.

Table 1. The situations of the final priority values with the proposed components.

$C(V_{GP})$	$C(WIQ)$	$C(component_1)$	$C(Priority)$
=	<	<	<
	=	=	=
	>	>	>
>	<	<	?
	=	=	>
	>	>	>
<	<	<	<
	=	=	<
	>	>	?

Overall, the proposed priority function with the above two components is expected to consider the effect of both the workload of machines (in component 1) and the other features (in component 2), and combine them together properly by multiplication.

4   Experiment Design

This work applies the framework (i.e. cooperative evolution genetic programming, CCGP) in [15], which is the current state-of-the-art algorithm of DFJSS to evolve routing and sequencing rules simultaneously. The proposed algorithm, which is named as rCCGP, is compared with CCGP [15] to verify its effectiveness on different job shop scenarios using three commonly used objectives, namely: (1) max-flowtime, (2) mean-flowtime, and (3) mean-weighted-flowtime.

To verify the performance of the evolved rules, we will use the test beds based on dynamic flexible simulation model [23,24]. In order to test the effectiveness and robustness of proposed algorithm, six simulation scenarios based on the three objectives and two utilisation levels (3 * 2) are investigated.

4.1   Simulation Configuration

For dynamic simulation, commonly used configuration is adopted. In the job shop, there are ten machines, which has been proven to be a good showcase for job shop environment. There are 5000 jobs that need to be processed by ten machines. In order to get a steady state, a warm up period of 1000 jobs is used and we collect data from the next 5000 jobs. The new jobs keep coming until the 6000th job is finished. In each problem instance, jobs arrive stochastically according to a Poison process with rate λ and the average processing time for machines has mean μ. The utilisation is the proportion of time (p) that a machine is busy as shown in Eq. (4). Two utilisation levels (i.e. 0.85 and 0.95) are used in this paper.

$$p = \lambda * \mu * P_M \tag{4}$$

In Eq. (4), P_M is the probability of a job visiting a machine. For example, if each job has two operations and there are ten machines, P_M is 2/10.

Different weights are set to jobs to indicate the urgency or importance of jobs (weight 1 (20%), weight 2 (60%), weight 4 (20%)). Uniform discrete distribution between 1 and 10 is designed for deciding both the number of operations per job and the number of candidate machines per operation. In addition, processing time of each operation will follow uniform discrete distribution between 1 and 99. In this work, the processing time of each operation is the same for all the candidate machines.

4.2   Parameter Settings

In our experiment, the terminal set and function set in [24] are adopted. The details are shown in Table 2. It is worth mentioned that "/" is the protected division that returns 1 if divided by 0.

Table 2. The terminal and function sets.

Terminals	Description
NIQ	The number of operations in the queue
WIQ	Current work in the queue
MWT	Waiting time of a machine
PT	Processing time of an operation
NPT	Median processing time for next operation
OWT	The waiting time of an operation
WKR	Median amount of work remaining of a job
NOR	The number of operations remaining of a job
W	Weight of a job
TIS	Time in system
Functions	$+, -, *, /, max, min$

The GP parameter settings follow the standard setting that have been used in most existing studies [15, 20, 22]. The population size is 1024 and the maximum depth of programs is 8. The crossover, mutation and reproduction rates are 0.80, 0.15 and 0.05, respectively. The rates of terminal and non-terminal selection are 0.10 and 0.90. Tournament selection is set as parent selection method with a tournament size of 7. The learning process continues until the generation reaches the maximum number of generations, which is set to 51.

5   Results and Discussions

The proposed GP approach with respect to the test performance and distribution of average objective value is investigated. 50 independent runs are executed, which assures that the results represent the average behavior instead of extreme situations. Then, the evolved rules are analysed. The $(-, +)$ marks show whether our proposed approaches converge significantly better or poorer than the basic approach in Wilcoxon rank sum test with a significance level of 0.05. Better results of min and max values are shown in bold. For the convenience of description, <obj, uti> indicates the simulation scenarios, where *obj* and *uti* are the objective and the utilisation level.

5.1   Test Performance of Evolved Rules

The comparison of the performance of the evolved rules obtained by rCCGP and CCGP are shown in Table 3. The statistical tests show that the evolved rules obtained by rCCGP are significantly better than the rules obtained by CCGP in four (scenario <Tmax, 0.95>, <Tmean, 0.85>, <WTmean, 0.85> and <WTmean, 0.95>) out of six scenarios. In scenario <Tmax, 0.85>, the performance is quite similar between rCCGP and CCGP with respective to the mean(sd), min and max value. In scenario <Tmean, 0.95>, although the rules evolved by rCCGP is not significantly better, the min, mean(sd) and max value are all better than their counterparts. In addition, rCCGP can reach better best-case and worse worst-case performance than CCGP in most cases.

Table 3. The mean(standard error), min and max of the objective value of rCCGP and CCGP over 50 independent runs for six dynamic scenarios.

Scenario	mean(sd)		min		max	
	rCCGP	CCGP	rCCGP	CCGP	rCCGP	CCGP
<Tmax,0.85>	1202.96(28.34)	1202.36(30.98)	1158.79	1152.97	1273.17	1270.26
<Tmax,0.95>	1864.83(30.69)(-)	1883.66(36.67)	**1813.77**	1829.47	**1942.71**	2034.34
<Tmean,0.85>	384.36(2.28)(-)	385.81(2.58)	**382.31**	382.72	396.83	395.81
<Tmean,0.95>	550.32(4.80)	552.14(6.46)	**543.18**	545.64	**569.43**	577.49
<WTmean,0.85>	828.30(6.02)(-)	829.38(3.47)	**823.08**	824.68	856.29	839.67
<WTmean,0.95>	1107.63(12.47)(-)	1110.72(10.77)	**1095.88**	1097.74	1169.34	1143.34

5.2   Distribution of Average Objective Value

Figure 2 shows the violin plot of the average objective value obtained by rCCGP and CCGP. When further looking into the violin plot in Fig. 2, we can see that in most scenarios, the value obtained by rCCGP are distributed at lower positions compared with the value achieved by CCGP expect for scenario <Tmax, 0.85>. Although in <Tmean, 0.85>, <WTmean, 0.85> and <WTmean, 0.95>, there are some outliers which are higher than the maximum outlier in CCGP, the number of outliers is still smaller than their counterparts. Except for the outliers, the value obtained by rCCGP are more concentrated than that of CCGP, even in scenario <Tmax, 0.85>. This suggests that the performance of GP with the proposed new representation is more stable and effective.

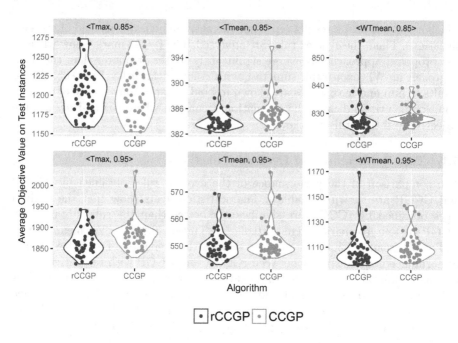

Fig. 2. Violin plot of average objective value obtained by rCCGP and CCGP.

5.3   Rule Analyses

Routing Rule. The proposed strategy for GP approach only works on routing process directly. The results show that the number of occurrences of the feature WIQ appeared in the final routing rules evolved by rCCGP is much lower than that of evolved by CCGP. For instance, in scenario <WTmean, 0.95>, the number of occurrences for feature WIQ in 50 best routing rules of CCGP (131) is 41% lighter than in rCCGP (93). This is consistent with our expectation as in the proposed rCCGP, the workload of machines is considered in component 1, so that the GP-evolved component 2 can be more focused on the other features.

Figure 3 shows the simplified component 2 of one promising routing evolved by rCCGP in the scenario <Tmax, 0.95>. It obtains a Tmax of 1831. There is no WIQ in the GP evolved rule (component 2). The component 2 consists of four parts and these four parts are added together. So, we can analyses them one by one. The details are shown in Eqs. (5)–(8). Note that based on the definition in Table 2, PT and NPT for all the candidate machines of an operation are equal in our experiment. NOR for a job is also the same for routing at the decision point. In addition, TIS and WKR are equal for an ready operation. In summary, by definition given in Table 2, at any routing decision situation, the PT, NPT, NOR, TIS and WKR features can be treated as constants, as they are the same for all the candidate machines. Adding or subtracting a constant can be removed from the priority function, since it will not change the relative preference between machines. In our experiment, the smaller the priority value, the higher the priority.

$$part_1 = 1 + max\{min\{NIQ, MWT\}, PT - MWT\} \\ - min(W, MWT + NOR) \tag{5}$$

The first part is shown as Eq. (5). Obviously, 1 and W are small constants, and thus $min(W, MWT + NOR)$ equals W in most cases since W is usually smaller than $MWT + NOR$. Therefore, Eq. (5) can be further simplified as $max\{min\{NIQ, MWT\}, PT - MWT\}$. It shows that the routing rule prefers machines with larger MWT, i.e. the earlier available machine (MWT = current time - machine ready time).

$$part_2 = PT * (1 + NIQ) - 2 * MWT - NOR + min\{NPT, TIS\} \tag{6}$$

The second part can be presented as Eq. (6). After ignoring some variables that can be considered as constants, Eq. (6) can be further simplified to $PT * NIQ - 2 * MWT$. It means the routing rule prefers machines with smaller NIQ (i.e. number of operations in the queue) and larger MWT.

$$part_3 = NPT + WKR \\ - min\{NIQ * PT, max\{min(NIQ, MWT), PT - MWT\}\} \tag{7}$$

The third part is described as Eq. (7). It can be simplified as $-min\{NIQ * PT, max\{min(NIQ, MWT), PT - MWT\}$ after ignoring the first two constant terms. In addition, no matter what $min\{NIQ * PT, max\{min(NIQ, MWT), PT - MWT\}$ returns, it will be cancelled out by the same component in $part_1$ or $part_2$.

$$part_4 = \frac{MWT * (PT + NPT + min\{NIQ, MWT\})}{OWT * W} \tag{8}$$

The last part can be denoted as Eq. (8). OWT (i.e. the waiting time of an operation) for a ready operation in routing process equals zero in our experiment (i.e. the details are shown in Sect. 2.2), therefore, the $part_4$ will return 1 (i.e. protected division).

According to the analysis mentioned above, this routing rule can be roughly simplified as $max\{min\{NIQ, MWT\}, PT - MWT\} - 2 * MWT$ or $PT * NIQ - 2 * MWT$.

Table 4 shows the number of times (proportion) a feature appears in the routing rule mentioned above and the counterpart in scenario <Tmax, 0.95>. The number of occurrences of features in the routing rule evolve by rCCGP and CCGP are 39 and 25, respectively. The number of designed terminals is 10 and the details can be seen in Table 2. The number of considered features in rCCGP (8) is more than that of evolved by CCGP (6). It suggests that the proposed GP approach can pay more attention to other features to get more information, thus to improve its performance. It is worth mentioning that the number of features considered in rCCGP is nine in fact because we consider WIQ by component 1.

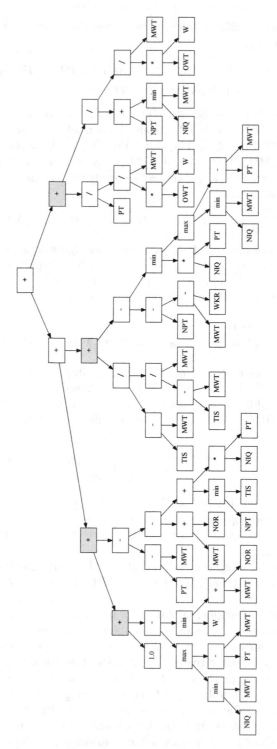

Fig. 3. Component 2 of one of the best performing rules for routing evolved by rCCGP in the scenario <tmax, 0.95>.

Table 4. The number of occurrences (proportion) of features in one promising routing rule evolved by rCCGP (component 2) and CCGP in scenario <Tmax,0.95>.

Feature	Count (rCCGP)	Count (CCGP)
MWT	15 (0.38)	7 (0.28)
PT	6 (0.15)	0(0.00)
NIQ	5 (0.13)	6 (0.24)
NPT	3 (0.08)	0 (0.00)
TIS	3 (0.08)	2 (0.08)
W	3 (0.08)	0 (0.00)
NOR	2 (0.05)	2 (0.08)
OWT	2 (0.05)	5 (0.20)
WKR	0 (0.00)	0 (0.00)
WIQ	0 (0.00)	3 (0.12)
Total	39	25

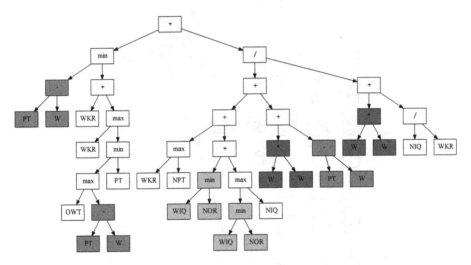

Fig. 4. Component 2 of one of the best performing rules for routing evolved by rCCGP in the scenario <tmax, 0.95>.

MWT, which occurs 15 times, is the most frequently seen feature $(15/39 = 0.38)$ of the evolved rule by rCCGP. For CCGP, MWT is also the most popularly used one, however, in terms of the proportion, it is less considered than that of in rCCGP $(0.28 < 0.38)$. rCCGP and CCGP pay different attention to different features (i.e. the features have different importance in rCCGP and CCGP).

Sequencing Rule. The corresponding sequencing rule of the routing rule compared in last section is observed here. This is mainly to investigate what effect

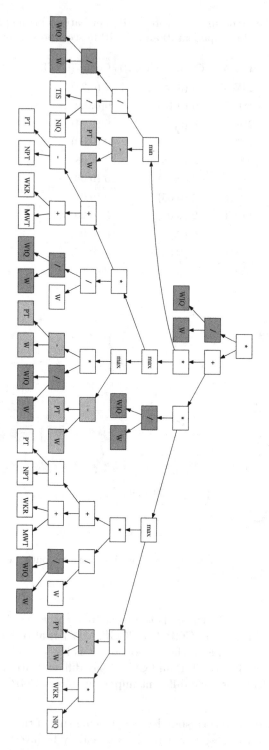

Fig. 5. The corresponding sequencing rule of the routing rule mentioned in last section evolved by CCGP in the scenario <tmax, 0.95>.

routing rule will have on sequencing rule. The sequencing rule evolved by rCCGP and CCGP are shown in Figs. 4 and 5. The size (i.e. number of nodes) of the sequencing rule is 45 evolved by rCCGP while the sequencing rule evolved by CCGP in the same scenario is 67. Obviously, the sequencing is much smaller.

When looking at the sequencing rule evolved by rCCGP, the most popular pattern is $PT - W$ followed by $W * W$ and $min\{WIQ, NOR\}$. WIQ/W appears most often (i.e. six times) followed by $PT - W$ in the sequencing rule evolved by CCGP. It means if the workload is not well considered in routing process, the machine might be assigned too many tasks. Thus, the sequencing rule should take WIQ into account. Intuitively, a machine with lots of work should take different processing strategy compared with a machine has fewer tasks.

6 Conclusions and Future Work

The goal of this paper was to help GP evolve more effective dispatching rules for DFJSS. The goal has been successfully achieved by developing a new representation. The new representation is based on the domain knowledge that the workload of a machine WIQ is the dominating feature for making the routing decision. To fully explore the best way of using the other non-dominating features, the new representation was designed as a combination of two parts, one solely related to the workload of machines, and the other focused on the other non-dominating features. This way, GP can focus on exploring more appropriate ways of using the other features than the dominating feature.

The results show that the proposed GP approach with new representation can achieve significantly better performance in most of the involved scenarios. To be specific, the distributions of average objective values obtained by rCCGP in five out of six scenarios are better than that of in CCGP. This means the proposed new representation works well in almost all the examined instances. It confirms the effectiveness of the proposed component for workload information and the combination strategy. It is also known that the routing rule obtained by rCCGP can focus more on the non-dominating features with the proposed new representation. In addition, the evolved corresponding sequencing rule is also affected to consider different information and tended to be smaller than its counterpart. Overall, the results demonstrate that the proposed way of using domain knowledge successfully helps GP evolve more effective routing and sequencing rules for DFJSS.

In the future, more strategies will be investigated to make full use of the information provided by features. In addition, useful techniques will be adopted to inspect the evolved rules.

References

1. Nguyen, S.B.S., Zhang, M.: A hybrid discrete particle swarm optimisation method for grid computation scheduling. In: 2014 IEEE Congress on Evolutionary Computation (CEC), pp. 483–490. IEEE (2014)

2. Tay, J.C., Ho, N.B.: Evolving dispatching rules using genetic programming for solving multi-objective flexible job-shop problems. Comput. Ind. Eng. **54**(3), 453–473 (2008)
3. Brucker, P., Schlie, R.: Job-shop scheduling with multi-purpose machines. Computing **45**(4), 369–375 (1990)
4. Oktaviandri, M., Hassan, A., Shaharoun, A.M.: Decision support tool for job shop scheduling with job cancellation. In: International Conference on Engineering of Taru-managara (ICET) (2013)
5. Sabuncuoglu, I., Bayız, M.: Analysis of reactive scheduling problems in a job shop environment. Eur. J. Oper. Res. **126**(3), 567–586 (2000)
6. Bertsekas, D.P., Bertsekas, D.P., Bertsekas, D.P., Bertsekas, D.P.: Dynamic Programming and Optimal Control, vol. 1. Athena Scientific, Belmont (2005)
7. Lawler, E.L., Wood, D.E.: Branch-and-bound methods: a survey. Oper. Res. **14**(4), 699–719 (1966)
8. Nowicki, E., Smutnicki, C.: A fast taboo search algorithm for the job shop problem. Manage. Sci. **42**(6), 797–813 (1996)
9. Pezzella, F., Morganti, G., Ciaschetti, G.: A genetic algorithm for the flexible job-shop scheduling problem. Comput. & OR **35**(10), 3202–3212 (2008)
10. Gomes, M.C., Barbosa-Póvoa, A.P., Novais, A.Q.: Reactive scheduling in a make-to-order flexible job shop with re-entrant process and assembly: a mathematical programming approach. Int. J. Prod. Res. **51**(17), 5120–5141 (2013)
11. Hildebrandt, T., Heger, J., Scholz-Reiter, B.: Towards improved dispatching rules for complex shop floor scenarios: a genetic programming approach. In: Proceedings of the 12th Annual Conference on Genetic and Evolutionary Computation, pp. 257–264. ACM (2010)
12. Pickardt, C.W., Hildebrandt, T., Branke, J., Heger, J., Scholz-Reiter, B.: Evolutionary generation of dispatching rule sets for complex dynamic scheduling problems. Int. J. Prod. Econ. **145**(1), 67–77 (2013)
13. Nguyen, S., Zhang, M., Johnston, M., Tan, K.C.: A computational study of representations in genetic programming to evolve dispatching rules for the job shop scheduling problem. IEEE Trans. Evol. Comput. **17**(5), 621–639 (2013)
14. Zhang, F., Mei, Y., Zhang, M.: Genetic programming with multi-tree representation for dynamic flexible job shop scheduling. In: Mitrovic, T., Xue, B., Li, X. (eds.) AI 2018. LNCS (LNAI), vol. 11320, pp. 472–484. Springer, Cham (2018). https://doi.org/10.1007/978-3-030-03991-2_43
15. Yska, D., Mei, Y., Zhang, M.: Genetic programming hyper-heuristic with cooperative coevolution for dynamic flexible job shop scheduling. In: Castelli, M., Sekanina, L., Zhang, M., Cagnoni, S., García-Sánchez, P. (eds.) EuroGP 2018. LNCS, vol. 10781, pp. 306–321. Springer, Cham (2018). https://doi.org/10.1007/978-3-319-77553-1_19
16. Zhang, F., Mei, Y., Zhang, M.: Surrogate-assisted genetic programming for dynamic flexible job shop scheduling. In: Mitrovic, T., Xue, B., Li, X. (eds.) AI 2018. LNCS (LNAI), vol. 11320, pp. 766–772. Springer, Cham (2018). https://doi.org/10.1007/978-3-030-03991-2_69
17. Haupt, R.: A survey of priority rule-based scheduling. Oper.-Res.-Spektrum **11**(1), 3–16 (1989)
18. Branke, J., Nguyen, S., Pickardt, C.W., Zhang, M.: Automated design of production scheduling heuristics: a review. IEEE Trans. Evol. Comput. **20**(1), 110–124 (2016)
19. Nguyen, S., Mei, Y., Zhang, M.: Genetic programming for production scheduling: a survey with a unified framework. Complex Intell. Syst. **3**(1), 41–66 (2017)

20. Mei, Y., Nguyen, S., Xue, B., Zhang, M.: An efficient feature selection algorithm for evolving job shop scheduling rules with genetic programming. IEEE Trans. Emerg. Top. Comput. Intell. **1**(5), 339–353 (2017)
21. Nguyen, S., Mei, Y., Xue, B., Zhang, M.: A hybrid genetic programming algorithm for automated design of dispatching rules. Evol. Comput. 1–31 (2018)
22. Mei, Y., Zhang, M., Nyugen, S.: Feature selection in evolving job shop dispatching rules with genetic programming. In: Proceedings of the Genetic and Evolutionary Computation Conference 2016, pp. 365–372. ACM (2016)
23. Hildebrandt, T., Branke, J.: On using surrogates with genetic programming. Evol. Comput. **23**(3), 343–367 (2015)
24. Mei, Y., Nguyen, S., Zhang, M.: Evolving time-invariant dispatching rules in job shop scheduling with genetic programming. In: McDermott, J., Castelli, M., Sekanina, L., Haasdijk, E., García-Sánchez, P. (eds.) EuroGP 2017. LNCS, vol. 10196, pp. 147–163. Springer, Cham (2017). https://doi.org/10.1007/978-3-319-55696-3_10

An Iterated Local Search Algorithm for the Two-Machine Flow Shop Problem with Buffers and Constant Processing Times on One Machine

Hoang Thanh Le[(✉)] [ID], Philine Geser, and Martin Middendorf [ID]

Swarm Intelligence and Complex Systems Group, Institute of Computer Science,
University of Leipzig, Leipzig, Germany
{lht,middendorf}@informatik.uni-leipzig.de

Abstract. This paper considers a special case of two-machine flow shop scheduling problems with buffers, namely, the case where all processing times on one of the two machines are equal. This case is interesting because it occurs in various applications, e.g., when one machine is a packing machine. For the buffers we consider two types of buffers that have been studied in the literature for flow shops. It is shown that all considered buffered flow shop problems remain *NP*-hard for the makespan criterion and permutation schedules even with the restriction to equal processing times on one machine. Two specific heuristics for solving the problems are proposed: (i) a modification of the commonly used NEH heuristic (mNEH) and (ii) an Iterated Local Search heuristic (2BF-ILS) that uses the mNEH heuristic for computing its initial solution. It is shown experimentally that the proposed 2BF-ILS heuristic obtains better results than two state-of-the-art algorithms for buffered flow shop problems from the literature and an Ant Colony Optimization algorithm. In addition, it is shown experimentally that 2BF-ILS can obtain the same solution quality as the standard NEH heuristic with a smaller number of function evaluations.

Keywords: Flow shops with buffers · Iterated Local Search ·
NEH heuristic · Permutation schedules

1 Introduction

Flow shop problems are a large class of job scheduling problems where each job (or object) goes through a set of m processing steps such that the ith processing step is done on machine M_i for $i \in \{1, \ldots, m\}$. The objective is to schedule the processing of jobs such that certain constraints are satisfied and certain criteria relating to efficiency, job throughput or incurred costs are optimized. In this paper we restrict ourselves to permutation schedules where the sequence of jobs is the same on all machines.

© Springer Nature Switzerland AG 2019
A. Liefooghe and L. Paquete (Eds.): EvoCOP 2019, LNCS 11452, pp. 50–65, 2019.
https://doi.org/10.1007/978-3-030-16711-0_4

The flow shop problems that are studied here are special cases of a two-machine flow shop with buffer constraints. In these problems each job takes up a given amount of buffer space for a certain time period and the scheduling has to be done such that the available buffer capacity is not exceeded. Buffer constraints occur in many practical applications where not only the time, but also the size of the processed objects has to be taken into account as, e.g., in production facilities with limited space for temporary storage. In particular, we consider two types of buffer constraints that have been studied for flow shops in the literature and which differ by the time span where the buffer is occupied by a job. Flow shops of these types occur, e.g., in waste processing or the production of metal parts for the aerospace industry [17], the scheduling of deliveries [2], the food industry or in the production of chemicals [14].

In this paper we consider the special case of buffered two-machine flow shops where for one of the two machines all jobs have the same processing time. This case is interesting because there exist many applications where in one stage of the flow shop the processing steps tend to be relatively similar for all objects. One example is a two-machine flow shop were different objects are produced on the first machine and packed on the second machine. In many applications of this type the production times might vary but packing is a standardized process which takes approximately the same time for each object. Another example is an application where a robot—which corresponds to the first machine—transports the materials for the production of different objects to the second machine where the manufacturing takes place. In many applications of this type the transportation times might be similar, e.g., when all materials are stored in the same storage hall, whereas the production times might differ strongly. In other applications of this type the opposite might hold: the transportation times are different whereas all production times are similar.

It is shown that the considered buffered two-machine flow shops remain *NP*-hard for the makespan criterion and permutation schedules even under the restriction to equal processing times on one machine. To solve the problems we propose a modification of the well-known NEH heuristic. Moreover, an Iterated Local Search heuristic (2BF-ILS) is proposed that uses the modified NEH heuristic for computing its initial solution. 2BF-ILS is studied experimentally and is compared with two state-of-the-art heuristics for buffered flow shops from the literature and also with an Ant Colony Optimization algorithm. 2BF-ILS is also compared experimentally to the standard NEH heuristic.

A short overview of recent works on buffer-constrained flow shop problems is given in Sect. 2. A formal description of the considered buffered flow shop problems is presented in Sect. 3. The *NP*-hardness results are shown in Sect. 4. The modified NEH heuristic is described in Sect. 5 and the Iterated Local Search heuristic is introduced in Sect. 6. Experimental results are presented in Sect. 7. Conclusions are given in Sect. 8.

2   Related Work

In the most often considered type of flow shop problem with buffers, there exists a buffer between each two adjacent machines M_i and M_{i+1} and each job

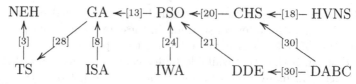

NEH: Nawaz-Enscore-Ham heuristic
GA: Genetic Algorithms
PSO: Particle Swarm Optimization
CHS: Chaotic Harmony Search
HVNS: Hybrid Variable Neighborhood Search

TS: Tabu Search
ISA: Immune System Algorithm
IWA: Invasive Weed Algorithm
DDE: Discrete Differential Evolution
DABC: Discrete Artificial Bee Colony

Fig. 1. Overview of works containing comparisons between algorithms. The notation A ← B indicates that algorithm A is outperformed by algorithm B in the given reference. However, the arrows are to be interpreted with some care: It does not mean that A is worse than B for two-machine flow shops with buffer constraints. It only means that B obtained better results (on average) than A for the considered test instances in the respective work.

occupies the buffer after it finishes on M_i and before it starts on machine M_{i+1}. This type of buffer is called *intermediate buffer*. It is known that the flow shop problem with intermediate buffers is *NP*-complete [22] and several authors have proposed heuristics for this problem, including a Tabu Search [11], Variable Neighborhood Search [18], Genetic Algorithms [28], methods based on Differential Evolution [21] and Particle Swarm Optimization [13]. Other examples of biologically inspired algorithms that have been applied on flow shop problems are a Discrete Artificial Bee Colony [30] and Immune System algorithms [1,8]. In most of these studies the performance of different heuristics has been compared. Figure 1 summarizes the results of these comparisons.

Another type of buffer-constraint that is considered in the literature is where the buffer between machines M_i and M_{i+1} is used by each job from its starting time on M_i until its finishing time on M_{i+1}. This type of buffer is referred to as *spanning buffer*. The flow shop problem with spanning buffer is also *NP*-complete [12]. Examples for methods from the literature for this type of flow shop are a Variable Neighborhood Search [10] and a Branch and Bound algorithm [12].

3   Formal Description of the Problem

For the two-machine flow shop problems (2-flow shop) considered in this paper there is a given set $\mathcal{J} = \{J_1, \ldots, J_n\}$ of n *jobs* and two *machines* M_1 and M_2. Each job J_i has two processing times $a_i > 0$, $b_i > 0$ where a_i is the processing time on M_1 and b_i is the processing time on M_2. In a 2-flow shop each job has to be finished on machine M_1 before it can be started on machine M_2. The processing of a job on a machine is non-preemptive and a machine can process at most one job at a time. A schedule for a 2-flow shop is denoted by σ and specifies

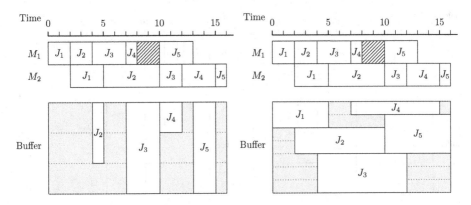

Fig. 2. Examples of a bi-2-flow-shop (left) with buffer capacity $\Omega = 3$ and a bs-2-flow-shop (right) with buffer capacity $\Omega = 7$ and $s_i = a_i$ for $i \in \{1, \ldots, n\}$. In each case a valid permutation schedule σ is shown with $\pi(\sigma) = (J_1, J_2, J_3, J_4, J_5)$. Blocks with diagonal lines indicate phases where M_1 is idle and the next job has to wait until enough buffer capacity becomes available.

the *starting* time $S_i^j(\sigma)$ and the *completion time* $C_i^j(\sigma)$ of J_i on M_j for each job J_i and each machine M_j ($i \in \{1, \ldots, n\}$, $j \in \{1, 2\}$). Thus, a schedule σ specifies *permutations* $\pi^1(\sigma), \pi^2(\sigma)$ describing the order of jobs on the machines M_1 and M_2, respectively. A schedule σ is a *permutation schedule* if $\pi^1(\sigma) = \pi^2(\sigma)$.

For the 2-flow-shop with buffer (b-2-flow-shop) there exists a buffer with a given capacity $\Omega \geq 0$ and each job $J_i \in \mathcal{J}$ has a size s_i which specifies how many buffer units J_i occupies when stored in the buffer. Two types of buffers are considered in this paper which differ by the time span in which a job occupies the buffer for a given schedule σ: (i) an *intermediate buffer* is occupied by job J_i from $C_i^1(\sigma)$ until $S_i^2(\sigma)$ and (ii) a *spanning buffer* is occupied by J_i from $S_i^1(\sigma)$ until $C_i^2(\sigma)$. The two types of b-2-flow-shops are denoted by bi-2-flow-shop and bs-2-flow-shop, respectively. A schedule σ is *valid* for a b-2-flow-shop when at every time step the total amount of buffer that is occupied by the jobs does not exceed Ω. If we impose the condition that all jobs are processed as early as possible, it is sufficient to specify the permutations $\pi^1(\sigma)$ and $\pi^2(\sigma)$ in order to define a schedule σ. Figure 2 shows examples of a bi-2-flow-shop (left) and a bs-2-flow-shop (right) together with a valid permutation schedule.

In this paper we concentrate on the special case of bs-2-flow-shop and bi-2-flow-shops where the processing times on one of the machines are the same for all jobs, i.e., it holds that $a_i = c$ or $b_i = c$ for $i \in \{1, \ldots, n\}$ with a constant c. With respect to the buffer usage of the jobs, two common models are studied in the literature: (i) the buffer is interpreted as a "counter" for the number of stored jobs and thus $s_i = 1$ for $i \in \{1, \ldots, n\}$ (see, e.g., [18,30]), and (ii) the amount of storage that is used by a job equals its processing time on the first machine, i.e., $s_i = a_i$ for $i \in \{1, \ldots, n\}$ (see, e.g., [6,10,12]). In the second model the size of the job J_i is interpreted as being proportional to the amount of work during its initial processing step.

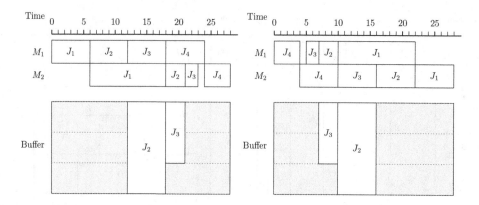

Fig. 3. Example for "mirroring" schedules in order to obtain the case $b_i = c$, $s_i = a_i$

Depending on whether the buffer usage s_i equals a_i, b_i, or 1 and whether the processing times on $M1$ or $M2$ are equal, it is possible to distinguish between six different cases. However, it can be shown that only two cases have to be considered since all other cases can be reduced to them. These two cases are (i) $b_i = c$, $s_i = a_i$ and (ii) $b_i = c, s_i = 1$. The reduction can be done by rescaling all values by s_i (if s_i is constant for all jobs) and/or by defining a new schedule σ' with the same makespan that "mirrors" the processing times (i.e., $a'_i = b_i$, $b'_i = a_i$ and $s'_i = s_i$) as well as the time axis (see Fig. 3 for an example).

The criterion to be optimized is the makespan $C_{\max}$ (i.e., the time where the last job is completed on the last machine). This is also the criterion used by the studies mentioned in Sect. 2 and based on these works, we also restrict ourselves to permutation schedules where the sequence of jobs on both machines is the same. Hence, a schedule is determined by a single permutation π of all jobs. However, it should be noted that this imposes a restriction for the cases with $s_i = a_i$, i.e., there exist instances where the set of optimal schedules contains no permutation schedule (for example, the bi-2-flow-shop in Fig. 3 right). For the case $s_i = 1$, there always exists a permutation schedule that is optimal [7].

4   Complexity Results

In this section, we show that for the considered b-2-flow-shops the special case where all processing times on machine M_2 are the same is NP-complete for the makespan criterion. Due to space limitations, the proof is sketched only for the case with intermediate buffer and $s_i = a_i$. The NP-completeness for the other considered flow shops can be shown with similar arguments.

Theorem 1. *The decision problem if there exists a valid permutation schedule σ^* with $C_{\max}(\sigma^*) \leq L$ for a given integer L is NP-complete for the following type of flow shops:*

(i) bi-2-flow-shop with $s_i = a_i$ and $b_i = c$ for $c > 0$,

(ii) bs-2-flow-shop with $s_i = a_i$ and $b_i = c$ for $c > 0$,
(iii) bi-2-flow-shop with $s_i = 1$ and $b_i = c$ for $c > 0$,
(iv) bs-2-flow-shop with $s_i = 1$ and $b_i = c$ for $c > 0$.

Proof (sketch for (i)). The problem is in NP since it can be checked in polynomial time if a schedule σ^* is valid and satisfies $C_{\max}(\sigma^*) \leq L$. To show the NP-hardness consider an instance of 3PARTITION: given are integers $x_1, x_2, \ldots, x_{3m}$ and an integer B such that $B/4 < x_j < B/2$ for all j and $\sum_{j=1}^{3m} x_j = mB$. The question is if there exists a partition of $\{x_1, x_2, \ldots, x_{3m}\}$ into m subsets $S_1, S_2, \ldots, S_m$ such that each set S_k satisfies $|S_k| = 3$ and $\sum_{x \in S_k} x = B$. We denote the given instance from 3PARTITION as I_{3P}. The corresponding flow shop instance I_{F2} with intermediate buffers is constructed with $n = 4m + 1$ jobs which are defined as follows:

– Job g_0 with $a_{g_0} = B/4$,
– A set G of m jobs $g_1, g_2, \ldots, g_m$ with $a_{g_k} = 3B$ for $k \in \{1, 2, \ldots, m\}$, and
– A set H of $3m$ jobs $h_1, h_2, \ldots, h_{3m}$ with $a_{h_k} = x_k$ for $k \in \{1, 2, \ldots, 3m\}$.

In addition, it is set that $\Omega = 3B/4$, $c = B$, $L = 4mB + B + B/4$ and $s_i = a_i$ for all jobs. Now it is shown that I_{3P} has a solution if and only if I_{F2} has a schedule σ^* with $C_{\max}(\sigma^*) \leq L$.

"$\Leftarrow$": Due to the properties of the jobs and the maximum makespan L, it can be argued that any solution σ^* to I_{F2} must have a structure as shown in Fig. 4 where $H_0, H_1, \ldots, H_{m-1}$ are sets of jobs in H and all jobs in $G \cup \{g_0\}$ are processed in succession without any delay. No other idle times are allowed and there are no H-jobs at the beginning or at the end of the schedule. Thus, all H-jobs must be processed in the "gaps" formed by the jobs in G. It can then be shown that the set H_0 contains exactly three H-jobs which are the same for both machines and satisfy $\sum_{h \in H_0} a_h = B$ as well as $\sum_{h \in H_0} b_h = 3B$. The arguments can be repeated to show the same properties for H_i, $i \in \{1, 2, \ldots, m-1\}$. The sets H_i then allows to construct a corresponding solution for the instance I_{3P} of 3PARTITION.

"$\Rightarrow$": Given the subsets $S_1, S_2, \ldots, S_m$ satisfying $|S_k| = 3$ and $\sum_{x \in S_k} x = B$ for $k \in \{1, 2, \ldots, m\}$, it is possible to construct a schedule as shown in Fig. 4 where the H-jobs corresponding to the subsets are scheduled between the G-jobs. This leads to a schedule σ^* satisfying $C_{\max}(\sigma^*) = L$ such that the resulting schedule σ^* is a solution for I_{F2}.   □

5 A Modification of the NEH Heuristic

The NEH algorithm [19] is a very successful and the most famous heuristic for makespan minimization in permutation flow shop problems. It is an iterative heuristic that starts with an empty permutation building it up by iteratively inserting jobs (in decreasing order of the sums of their processing times) into the positions which lead to the smallest increase in makespan. Several variants

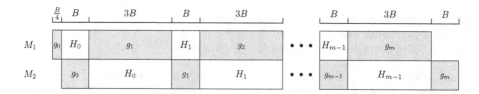

Fig. 4. Visualization for Theorem 1

of the NEH heuristic have been studied in the literature. Most of them deal with ties [5] or use different orders of job insertions, e.g., based on the moments of processing times [15] or using genetic programming [27]. A disadvantage of the NEH heuristic is its relatively large run time since it checks $\Theta(n^2)$ insertion points. In particular, when the solution of the NEH heuristic is further used by improvement heuristics, e.g., as the starting solution for an Iterated Local Search heuristic, it is desirable to have a faster heuristic for the computation of a starting solution.

Here, we introduce a modification of the NEH heuristic (mNEH) that is faster and particularly suitable for the considered two-machine flow shops with buffers where all processing times on the second machine M_2 are equal. In this case, the jobs differ only by their processing time on M_1. Therefore, it will often happen that many jobs are similar or even identical. This property is used in the mNEH heuristic to reduce the number of positions that are checked for the insertions of the jobs and thus the total number of evaluated schedules. The main idea is to split the n jobs into $G(n)$ groups of similar jobs and to maintain for each group a list of $L(n)$ "good" candidate positions for the insertion. This is based on the assumption that insertions of similar jobs at the same positions will lead to similar schedules.

Algorithm 1. Modified NEH heuristic (mNEH)

Parameters: number of groups $G(n)$, number of positions $L(n)$ to memorize

1: $\pi^{sort} \leftarrow$ sorted sequence of jobs by descending a_i
2: $\pi \leftarrow$ empty permutation
3: $S_1, S_2, \ldots, S_{G(n)} \leftarrow$ partition of π^{sort} into $G(n)$ groups of equal size
4: **for** $S \in \{S_1, S_2, \ldots, S_{G(n)-1}\}$ **do**
5:     $(j_1, j_2, \ldots, j_k) \leftarrow$ jobs in the current group S
6:     test insertion of j_1 in all possible positions of π and memorize the best $L(n)$ insertions
7:     insert j_1 at the best position in π
8:     update list of memorized $L(n)$ positions
9:     **for** $\ell \in \{2, 3, \ldots, k\}$ **do**
10:         test insertion of j_ℓ into $\hat{\pi}$ at the memorized $L(n)$ positions
11:         test insertion of j_ℓ into neighbor positions of $j_{\ell-1}$
12:         insert j_ℓ at the best tested position in π
13:         update list of memorized $L(n)$ positions, if necessary
14:     **end for**
15: **end for**
16: insert the jobs in $S_{G(n)}$ into π as in the standard NEH heuristic
17: **return** π

A pseudocode of the modified heuristic is given in Algorithm 1. The number of evaluations to be performed can then be adjusted by appropriate choice of $L(n)$ and $G(n)$. In this paper we memorize $L(n) = 2\sqrt{n}$ positions for each of $G(n) = \sqrt{n}$ groups of equal size (an exception is the last group as described later) and it follows that the resulting algorithm performs $O(n\sqrt{n})$ evaluations. It should be noted that in this case the incomplete permutation π contains less than $L(n)$ jobs during the scheduling of the first two groups S_1 and S_2. For the jobs in these two groups, the mNEH heuristic tests all possible insertion positions in the same way as in the standard NEH heuristic. For the last group $S_{G(n)}$, all possible positions are checked since this group contains the shortest jobs allowing for a finer optimization of the partial permutation π. Regarding the time complexity (including the time to evaluate a permutation as well as the time to build and update the list) it can be shown that the mNEH heuristic with the used parameter values is faster than the standard NEH heuristic by a factor of $O(\sqrt{n})$.

6 Iterated Local Search

In this section we propose an Iterated Local Search heuristic for the two-machine flow shop problems with buffers. The heuristic is denoted by 2BF-ILS. The local search operations that are considered for 2BF-ILS are the following:

- Inserting a job on position i into another position j (*insert*)
- Inserting a pair of adjacent jobs at the positions $(i, i+1)$ into the positions $(j, j+1)$ (*pairInsert*)
- Swapping two jobs at the positions i and j (*swap*)

The actual selection of these operations and their order used in the experiment is later determined by the algorithm configurator irace in Sect. 7.3. As such, the following description is based on the generalized case where a sequence $op_1, op_2, \ldots, op_\ell$ is given with $op_i \in \{insert, pairInsert, swap\}$ for $i \in \{1, \ldots, \ell\}$. Each of the considered operations $op \in \{insert, pairInsert, swap\}$ takes two parameters i and j so that we denote the result of op applied on π with these parameters as $op(\pi, i, j)$. The resulting neighborhood of permutations around π is denoted $N^{op}(\pi) = \{\pi' \mid \exists i, j : op(\pi, i, j) = \pi'\}$, i.e., the set of all permutations π' that can be obtained by a single application of op on π. The size of the neighborhoods for the considered operations is quadratic with respect to the problem size n since the number of possible values for the parameters i and j linearly increases with n. One way to reduce the size is to fix a parameter i:

$$N_i^{op}(\pi) = \{\pi' \mid \exists j : op(\pi, i, j) = \pi'\}$$

This lowers the number of checked permutations leading to potentially good solutions being missed. However, the linear size of the resulting neighborhood allows for more local search steps to be performed in the same amount of time.

Algorithm 2. 2BF-ILS

Parameters: initial perturbation strength ps_{init}, finite sequence of operations $(op_1, op_2, \ldots, op_\ell)$

1: $\pi^0 \leftarrow$ permutation generated by mNEH (Algorithm 1)
2: $ps \leftarrow ps_{init}$
3: **while** termination criterion not satisfied **do**
4:      $\pi^{cur} \leftarrow$ best known solution
5:      $\pi^{rand} \leftarrow$ random job permutation
6:      **if** best solution did not improve in previous iteration **then**
7:          $\pi^{cur} \leftarrow perturb(\pi^{cur}, ps)$
8:      **end if**
9:      **for** $s \in \{1, 2, \ldots, \ell\}$ **do**
10:         $op \leftarrow op_s$
11:         **repeat**   ▷ local search
12:             **for** $k \in \{1, 2, \ldots, n\}$ **do**
13:                 $i \leftarrow k$th element in π^{rand}
14:                 $\hat{\pi} \leftarrow$ best permutation in the neighborhood $N_i^{op}(\pi^{cur})$
15:                 **if** $C_{\max}(\hat{\pi}) \leq C_{\max}(\pi^{cur})$ **then**
16:                     $\pi^{cur} \leftarrow \hat{\pi}$
17:                 **end if**
18:             **end for**
19:         **until** π^{cur} does not improve
20:     **end for**
21:     **if** best known solution did not improve in current iteration **then**
22:         $ps \leftarrow ps + \varepsilon$
23:     **else**
24:         $ps \leftarrow ps_{init}$
25:     **end if**
26: **end while**
27: **return** best known solution

The Algorithm 2 outlines the main steps of our proposed method: It starts with the solution obtained from the mNEH heuristic proposed in Sect. 5 (with $G(n) = \sqrt{n}$ and $L(n) = 2\sqrt{n}$). Then 2BF-ILS iterates through a sequence of operations $(op_1, op_2, \ldots, op_\ell)$ and repeatedly performs local search steps with changing neighborhoods. The choice of which parameter i to fix in $N_i^{op}(\pi)$ is based on a random permutation π^{rand} calculated beforehand. The search steps with the current operation op are repeated until the permutation π^{cur} obtained so far cannot be improved further. Afterwards, the local search procedure is restarted with the next given operation.

If the best known solution obtained so far did not improve after all operations, the next iteration of the algorithm uses a perturbed variant of the best known solution. The perturbation we use is based on a geometric distribution: We repeatedly draw (uniformly) random numbers r from $[0, 1]$ and then randomly apply one of the given operations with random parameters on the permutation until r is greater than ps. This distribution favors a small number of perturbations. The strength of perturbation ps increases additively (here set as $\varepsilon = 0.05$ with ps maxed out at 0.99) if successive iterations do not yield any improvement of the best known solution. If it is improved, the value ps is reset to the initial value ps_{init}.

7 Computational Evaluation

In this section we describe the experiments and their results.

7.1 Choice of Algorithms for Comparison

Based on the literature overview given in Sect. 2 and Fig. 1, we selected the *Hybrid Variable Neighborhood Search* (HVNS) and the *Discrete Artificial Bee Colony* (DABC) for a comparison with our proposed method since these algorithms are described in fairly recent works [18,30] and have not been outperformed by any other algorithm to the best of our knowledge. For the implementation, we asked the authors of both algorithms for the source code but did not receive a reply. Thus, we reimplemented both algorithms.

In addition to HVNS and DABC we also used an *Ant Colony Optimization* algorithm (ACO) for comparison. The reason is that ACO generally achieves good results for many scheduling problems. It is thus a promising algorithm which so far has not been applied to flow shops with buffers. The ACO in this work is a standard ACO where each pheromone value τ_{ij} encodes the desirability that job J_j is the successor of job J_i. At each iteration the pheromone values are updated by the solution that was obtained by 2BF-ILS applied to the permutation generated by the best ant in the current iteration (i.e., the iteration best solution became π^{cur} in line 4 in Algorithm 2). Each application of 2BF-ILS was limited to 1000 search steps.

7.2 Generation of Problem Instances

The commonly used benchmark instances for flow shop problems (from Taillard [25], Reeves [23], Carlier [4] and the recent VRF benchmark [26]) work with at least five machines and contain no buffer constraints so that they cannot be directly applied to the b-2-flow shop cases considered here. Authors that studied buffered flow shop problems with two machines (e.g., in [10,12]) generated instances with random processing times uniformly drawn over the interval $\{1, \ldots, 100\}$. Based on the studies performed by the these authors, we generated instances for our experimental study as described in the following.

We chose $n \in \{50, 100, 150\}$ as the number of jobs for small, medium and large instances and created three "incomplete instances" for each size that contain only the M_1 processing times a_j drawn randomly from a uniform distribution over the set $\{1, 2, \ldots, 100\}$. Each of these incomplete instances was then used to build three subordinate instances by choosing the constant processing times c on M_2 as the 25%, 50% or the 75% percentile $q_{0.25}, q_{0.50}, q_{0.75}$ of the values a_i. Each value of c was then used to create (complete) instances for the intermediate buffer (or spanning buffer) by setting $s_i = 1$ and $\Omega = 1$ ($\Omega = 3$) or $s_i = a_i$ and $\Omega = q_{0.25}$ ($\Omega = \max a_i + q_{25}$). Note that the buffer capacity Ω cannot be smaller than $\max s_i$ for the spanning buffer model or else there exists no valid schedule. The resulting set of instances contained 108 flow shop problems.

7.3 Parameter Values

For the algorithms 2BF-ILS and ACO the parameter values were determined with the algorithm configurator `irace` [16]. The configuration was performed on a separate set of instances with 100 jobs (generated using the method described in Subsect. 7.2) with the standard `irace` parameter values and a budget of 250 runs for each algorithm. The limits used by `irace` for the parameter values were chosen as $ps_{init} \in (0.01, 0.99)$, $\rho \in (0.01, 0.05)$, $N_{ants} \in \{1, 2, \ldots, 20\}$. As possible sequences for operations we considered all sequences of length 1,2 or 3 that can be formed by the operations *insert*, *pairInsert* or *swap* (without repetition leading to 15 possible sequences in total). The algorithms HVNS and DABC were applied with the parameters given by the respective authors as well as tuned parameters calculated by `irace` on the same instances. In the following, the tuned versions of these algorithms are referred to as HVNS-T and DABC-T.

Table 1. Parameters for ACO and 2BF-ILS calculated by `irace`

Algorithm	Parameter	Result	Description
ACO	ps_{init}	0.914	Initial perturbation strength
	op	(*swap, insert, pairInsert*)	Sequence of operations
	N_{ants}	19	Number of ants
	ρ	0.018	Pheromone evaporation
2BF-ILS	ps_{init}	0.240	Initial perturbation strength
	op	(*pairInsert, insert, swap*)	Sequence of operations

The resulting parameter values are given in Table 1. Due to page limitations, we refer the reader to https://github.com/L-HT/TwoMachineFlowShopBuffers where the resulting parameter values for HVNS-T and DABC-T as well as the source code for each algorithm and the following evaluations (written in R and C++) are available. After the configuration, each algorithm was executed on all 108 instances and the resulting values were averaged over 10 replications. All test runs were performed on a computer with eight 3.4-GHz-cores (each run being executed on one core) and 32 GB RAM with time limits of 5, 10 and 15 min for the small, medium and large problems, respectively.

7.4 Comparison of 2BF-ILS with Other Metaheuristics

The performance of each algorithm was evaluated based on (i) the number of performed *function evaluations FE* to calculate the makespan and (ii) the elapsed absolute runtime. The latter was used to calculate the *normalized runtime NT* as a time measure where the reference runtime for a given instance was chosen as the mean runtime of 10 runs of the standard NEH heuristic.

For each instance and each time measure, we generated (similar to what is done in [29]) the *progress curve* (PC, the quality of the best known solution

Table 2. Relative area under curve (AUC) values for each algorithm, evaluation diagram and time measure (indicated by the subscript NT or FE). For each instance, the area under the curve was calculated for each diagram and the best obtained value was chosen as the "reference" to calculate the relative values for the other algorithms. The obtained values were averaged over all 108 instances. A low number indicates that the curves for the respective algorithm show a high similarity (on average) to the best performing algorithm on each instance.

	HVNS	HVNS-T	DABC	DABC-T	ACO	2BF-ILS
PC_{FE}	2.22	2.13	1.81	1.66	2.01	1.03
PC_{NT}	2.41	2.40	1.54	1.71	2.06	1.04
ERT_{FE}	2.62	2.59	3.07	3.02	2.58	1.11
ERT_{NT}	2.49	2.48	1.91	2.36	2.60	1.25
$ECDF_{FE}$	1.28	1.27	1.23	1.23	1.44	1.00
$ECDF_{NT}$	1.35	1.35	1.21	1.22	1.50	1.00

over time), the *estimated running time* (ERT, indicating the time necessary to reach a given solution quality) and the *empirical cumulative distribution function* (ECDF, describing the percentage of runs reaching a given solution quality over time) with respect to the relative percentage difference $RPD = (F(\pi) - \hat{F})/\hat{F}$ between the solution quality $F(\pi)$ of the permutation π and the best solution quality $\hat{F}$ found in all runs on this instance. The target value of the ECDF was chosen as a 1%-deviation of $\hat{F}$ such that these diagrams show how consistent an algorithm reaches the performance $RPD \leq 0.01$.

Since an individual evaluation of over 600 diagrams is not feasible, we calculated the *area under curve* (AUC) as an aggregate quality measure for the PC, ERT and ECDF diagrams (with respect to both time measures) since algorithms with low AUC values (for PC and ERT diagrams) and high AUC values (for ECDF diagrams) tend to find better solutions faster [29]. As such, these values quantify the performance of an algorithm for a given instance over time. The relative AUC values, averaged over all instances used in this study, are shown in Table 2 It can be seen that 2BF-ILS outperformed the other algorithms and their `irace`-tuned versions with respect to all evaluation measures. ACO shows a competitive performance when compared to HVNS and DABC with respect to the values for the PC diagrams which indicates that it quickly improves at the beginning when the first solutions are created. One reason why ACO does not perform very well on the studied type of buffered flow shop problems is that there exist many similar or equal jobs which causes a problem for the ants.

In addition, we observed for most instances with the makespan criterion and $s_i = 1$ (for both buffer types) that the algorithms quickly reached a state of convergence with the same solution quality (with an example shown in Fig. 5 left). A comparison with the lower bound given by Johnson's algorithm [9] for flow shops without buffer constraints showed that in most cases an optimal solution was reached. We performed further experiments using additional instances of this type with similar results which indicates that this special case could be "easier"

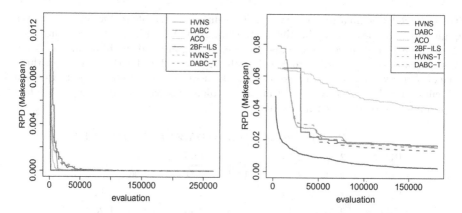

Fig. 5. Progress curve for an instance using the intermediate buffer, $s_i = 1$, $c = q_{0.50}$ and 50 jobs (left) and an instance with spanning buffer, $s_i = a_i$, $c = q_{0.50}$ and 100 jobs (right)

to solve than other problems (even though in theory it is still *NP*-complete). This was not observed for the instances with $s_j = a_j$ where a slower convergence occurred in most cases (see Fig. 5 right for an example).

Table 3. Results of the pairwise comparisons between the algorithms using the two-sided sign test ($n = 108$ instances). The first value in each cell shows the test result with respect to the performance at 100 000 evaluations and the second value refers to the performance reached at the end of the time limit. A triangle indicates that the measured difference is statistically significant ($p < 0.05/15$ due to Bonferroni correction) and that the algorithm at which the triangle is pointed at is significantly better according to the test statistic.

	HVNS	HVNS-T	DABC	DABC-T	ACO	2BF-ILS
HVNS		- / -	▲/▲	▲/▲	- / -	▲/▲
HVNS-T	- / -		- /▲	▲/▲	- / -	▲/▲
DABC	◄/◄	- /◄		- / -	- /◄	▲/▲
DABC-T	◄/◄	◄/◄	- / -		◄/◄	▲/▲
ACO	- / -	- / -	- /▲	▲/▲		▲/▲
2BF-ILS	◄/◄	◄/◄	◄/◄	◄/◄	◄/◄	

To compare the performance of the algorithms at specific points in time the sign test for paired samples was applied. This non-parametric test neither requires the given data to be normally distributed nor the difference distributions between the methods to be symmetric. Using this test, we compared the performance of the algorithms during the run (at 100 000 evaluations) and the performance that was reached at the end of the time limit. The results are shown in Table 3 for these two points in time. Note that this table is symmetric since the 15 possible pairwise comparisons for each of the two points in time were performed with two-sided

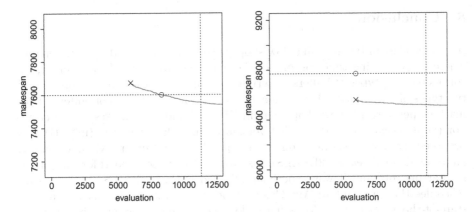

Fig. 6. Comparison between the standard NEH and 2BF-ILS for an instance with 150 jobs, spanning buffer, $s_i = 1$, $c = q_{0.50}$ (left) and an instance with 150 jobs, spanning buffer, $s_i = a_i$ and $c = q_{0.50}$ (right). Dashed lines indicate when the standard NEH finishes (vertical lines) and the attained solution quality (horizontal lines). A cross marks the quality of the initial mNEH solution and a circle marks the number of evaluations at which a better quality than NEH was reached for the first time.

tests. Similar to Table 2, it can be seen that 2BF-ILS obtained a significantly better performance than the other algorithms and their tuned versions at both points indicating a high and consistent performance over time.

7.5   Comparison of 2BF-ILS with NEH

In this section we compare the standard NEH with 2BF-ILS which uses mNEH with the parameter values $G(n) = \sqrt{n}$ and $L(n) = 2\sqrt{n}$ for constructing its starting solution. In particular, the time was measured for 2BF-ILS to obtain the same solution quality C_{NEH} as the standard NEH. This experiment was performed using the same instances described in Subsect. 7.2.

For an example instance, Fig. 6 shows the solution quality of the standard NEH heuristic as well as the performance over time for 2BF-ILS (averaged over 10 replications). In the left part of Fig. 6 it can be seen that mNEH only needed approximately half as many evaluations as NEH and initially obtained a slightly worse solution quality. However, the reduced number of evaluations allowed 2BF-ILS to obtain the quality C_{NEH} with fewer evaluations than NEH. This was observed for most problem instances and the sign test ($n = 108$) showed that 2BF-ILS needed significantly fewer evaluations to reach C_{NEH} than NEH ($p < 0.001$). For some instances with a spanning buffer and $s_i = a_i$, mNEH even obtained a better initial performance (e.g., Fig. 6 right). Even though the differences for these instances were not statistically significant ($n = 27$, $p = 0.08$), it is interesting to further investigate for which b-2-flow-shops mNEH performs particularly well.

8 Conclusion

A specific class of two-machine flow shops with buffers was studied in this paper where all processing times on one of the machines are equal. In particular, we considered two types of buffers (intermediate buffers and spanning buffers) and two types of buffer usage. It was shown that all considered cases of buffered two-machine permutation flow shops remain NP-hard under the makespan criterion. For these special cases a modified version of the NEH heuristic (mNEH) was proposed that uses fewer function evaluations than the standard NEH. Moreover, an Iterated Local Search (2BF-ILS) was proposed that uses mNEH for computing its starting solution. It was experimentally shown that 2BF-ILS obtains better schedules than a standard Ant Colony Optimization algorithm as well as two state-of-the-art algorithms for buffered flow shop problems from the literature. In addition, experimental results showed that 2BF-ILS obtains the same solution quality as the standard NEH heuristic but uses a smaller number of function evaluations.

Acknowledgements. This work was funded by the Deutsche Forschungsgemeinschaft (DFG, German Research Foundation) - project number 392050753.

References

1. Abdollahpour, S., Rezaeian, J.: Minimizing makespan for flow shop scheduling problem with intermediate buffers by using hybrid approach of artificial immune system. Appl. Soft Comput. **28**(C), 44–56 (2015)
2. Aloulou, M.A., Bouzaiene, A., Dridi, N., Vanderpooten, D.: A bicriteria two-machine flow-shop serial-batching scheduling problem with bounded batch size. J. Sched. **17**(1), 17–29 (2014)
3. Brucker, P., Heitmann, S., Hurink, J.: Flow-shop problems with intermediate buffers. Oper. Res. Spektrum **25**(4), 549–574 (2003)
4. Carlier, J.: Ordonnancements a contraintes disjonctives. R.A.I.R.O. Recherche operationelle **12**, 333–351 (1978)
5. Fernandez-Viagas, V., Framinan, J.M.: NEH-based heuristics for the permutation flowshop scheduling problem to minimise total tardiness. Comput. Oper. Res. **60**, 27–36 (2015)
6. Fung, J., Zinder, Y.: Permutation schedules for a two-machine flow shop with storage. Oper. Res. Lett. **44**(2), 153–157 (2016)
7. Geser, P.: Zur Komplexität des 2-Maschinen-Flow-Shop-Problems mit Zwischenspeicher. Diploma thesis, Leipzig University (2017)
8. Hsieh, Y.C., You, P.S., Liou, C.D.: A note of using effective immune based approach for the flow shop scheduling with buffers. Appl. Math. Comput. **215**(5), 1984–1989 (2009)
9. Johnson, S.M.: Optimal two- and three-stage production schedules with setup times included. Naval Res. Logist. Q. **1**(1), 61–68 (1954)
10. Kononova, P.A., Kochetov, Y.A.: The variable neighborhood search for the two machine flow shop problem with a passive prefetch. J. Appl. Ind. Math. **7**(1), 54–67 (2013)

11. Li, S., Tang, L.: A tabu search algorithm based on new block properties and speed-up method for permutation flow-shop with finite intermediate storage. J. Intell. Manuf. **16**(4), 463–477 (2005)
12. Lin, F.C., Hong, J.S., Lin, B.M.T.: A two-machine flowshop problem with processing time-dependent buffer constraints-an application in multimedia presentations. Comput. Oper. Res. **36**(4), 1158–1175 (2009)
13. Liu, B., Wang, L., Jin, Y.H.: An effective hybrid PSO-based algorithm for flow shop scheduling with limited buffers. Comput. Oper. Res. **35**(9), 2791–2806 (2008)
14. Liu, S.Q., Kozan, E.: Scheduling a flow-shop with combined buffer conditions. Int. J. Prod. Econ. **117**(2), 371–380 (2009)
15. Liu, W., Jin, Y., Price, M.: A new improved NEH heuristic for permutation flow-shop scheduling problems. Int. J. Prod. Econ. **193**, 21–30 (2017)
16. López-Ibáñez, M., Dubois-Lacoste, J., Pérez Cáceres, L., Stützle, T., Birattari, M.: The irace package: iterated racing for automatic algorithm configuration. Oper. Res. Perspect. **3**, 43–58 (2016)
17. Martinez, S., Dauzère-Pérès, S., Guéret, C., Mati, Y., Sauer, N.: Complexity of flowshop scheduling problems with a new blocking constraint. Eur. J. Oper. Res. **169**, 855–864 (2006)
18. Moslehi, G., Khorasanian, D.: A hybrid variable neighborhood search algorithm for solving the limited-buffer permutation flow shop scheduling problem with the makespan criterion. Comput. Oper. Res. **52**, 260–268 (2014)
19. Nawaz, M., Enscore, E.E., Ham, I.: A heuristic algorithm for the m-machine, n-job flow-shop sequencing problem. Omega **11**, 91–95 (1983)
20. Pan, Q.K., Wang, L., Gao, L.: A chaotic harmony search algorithm for the flow shop scheduling problem with limited buffers. Appl. Soft Comput. **11**, 5270–5280 (2011)
21. Pan, Q.K., Wang, L., Gao, L., Li, W.D.: An effective hybrid discrete differential evolution algorithm for the flow shop scheduling with intermediate buffers. Inf. Sci. **181**(3), 668–685 (2011)
22. Papadimitriou, C.H., Kanellakis, P.C.: Flowshop scheduling with limited temporary storage. J. Assoc. Comput. Mach. (JACM) **27**(3), 533–549 (1980)
23. Reeves, C.: A genetic algorithm for flowshop sequencing. Comput. Oper. Res. **22**(1), 5–13 (1995)
24. Sang, H.Y., Pan, Q.K.: An effective invasive weed optimization algorithm for the flow shop scheduling with intermediate buffers. In: 25th Chinese Control and Decision Conference (CCDC), pp. 861–864 (2013)
25. Taillard, E.: Benchmarks for basic scheduling problems. Eur. J. Oper. Res. **64**, 278–285 (1993)
26. Vallada, E., Ruiz, R., Framinan, J.M.: New hard benchmark for flowshop scheduling problems minimising makespan. Eur. J. Oper. Res. **240**, 666–677 (2015)
27. Vázquez-Rodríguez, J.A., Ochoa, G.: On the automatic discovery of variants of the neh procedure for flow shop scheduling using genetic programming. J. Oper. Res. Soc. **62**(2), 381–396 (2011)
28. Wang, L., Zhang, L., Zheng, D.Z.: An effective hybrid genetic algorithm for flow shop scheduling with limited buffers. Comput. Oper. Res. **33**, 2960–2971 (2006)
29. Weise, T., Chiong, R., Lassig, J., Tang, K., Tsutsui, S., Chen, W., Michalewicz, Z., Yao, X.: Benchmarking optimization algorithms: an open source framework for the traveling salesman problem. IEEE Comput. Intell. Mag. **9**(3), 40–52 (2014)
30. Zhang, S.J., Gu, X.S.: An effective discrete artificial bee colony algorithm for flow shop scheduling problem with intermediate buffers. J. Central South Univ. **22**(9), 3471–3484 (2015)

Route Planning for a Fleet of Electric Vehicles with Waiting Times at Charging Stations

Baoxiang Li[✉], Shashi Shekhar Jha, and Hoong Chuin Lau

Fujitsu-SMU Urban Computing and Engineering Corporate Lab,
School of Information Systems, Singapore Management University,
Singapore, Singapore
bxli@smu.edu.sg

Abstract. Electric Vehicles (EVs) are the next wave of technology in the transportation industry. EVs are increasingly becoming common for personal transport and pushing the boundaries to become the mainstream mode of transportation. Use of such EVs in logistic fleets for delivering customer goods is not far from becoming reality. However, managing such fleet of EVs bring new challenges in terms of battery capacities and charging infrastructure for efficient route planning. Researchers have addressed such issues considering different aspects of the EVs such as linear battery charging/discharging rate, fixed travel times, etc. In this paper, we address the issue of waiting times due to limited charging capacity at the charging stations while planning the routes of EVs for providing pickup/delivery services. We provide an exact mathematical model of the problem considering waiting times of vehicle based on their arrival at the charging stations. We further develop a genetic algorithm approach that embeds Constraint Programming to solve the problem. We test our approach on a set of benchmark Solomon instances.

Keywords: Electric Vehicle Routing Problem ·
Mixed integer linear programming · Constraint Programming ·
Genetic algorithm

1 Introduction

With rising pollution levels globally, cities are adopting means and technologies that are social and environmental friendly for various activities. The movement of vehicles on the roads is one of the major contributors of the overall rise in the air pollution over the globe. Recent advancements in the electric vehicle technology with high power and compact batteries have opened up new modalities to shape the transportation for people and goods. The prominent advantage of using electric vehicles is that they do not emit the green house gases that act as the major contributor to the global pollution. For large scale logistics operations such Electric Vehicles (EVs) now provide more viable green technology to serve customers (mostly referred as green logistics). However, EVs have their own

A. Liefooghe and L. Paquete (Eds.): EvoCOP 2019, LNCS 11452, pp. 66–82, 2019.
https://doi.org/10.1007/978-3-030-16711-0_5

share of drawbacks which limits their usage. At the very basic, it is the limited driving range with one complete charge that EVs can scale up to. The maximum driving range of the EVs is around 100–150 miles which gets further reduced by the operating and environmental conditions such as low temperatures, battery age, weight of the vehicle with payload, etc. Further EVs can take more than half-an-hour to completely charge their batteries which is also dependent on the type of batteries and the capacity of charging outlets.

To make way for the EVs to join the mainstream commercial logistics services, one needs to consider all such constraints for route planning and scheduling of these vehicles. Although researchers have considered the limited battery capacity of the EVs in planning the routes, the limited capacity of charging stations have scarcely been addressed in the literature. The scarcity of charging station resources is particularly pertinent in cities where EVs are still in the infant stage of adoption. The station locations, number of charging outlets and charge delivery capacities put limitations to the number of EVs that can be charged simultaneously at a charging station. In addition, due to the long charging time of EVs with limited charging infrastructure, one needs to consider the significant queuing time at the charging stations while planning routes for logistics fleet operations.

In this paper, our focus is to plan the routes for a fleet of EVs for logistics operations by allowing the EVs to queue and wait at the fixed capacity charging stations. We present an exact mathematical formulation of the classical VRP problem with the considerations of limited battery charge, cargo capacity constraints and waiting times at the charging stations. One major difficulty in modeling this problem is to handle the queuing sequence at each charging station, which we overcome by using a subtour elimination technique commonly used for solving the Traveling Salesman Problem (TSP). Such considerations can make the route planning of EVs more viable for real world operations. Further, the last mile delivery operations can further be streamlined with such EVs taking the central stage with carefully planned routes and visits to the charging stations. Hence, our work extends the current literature by addressing the challenging aspect of route planning for a logistics fleet with an explicit modeling of waiting times at the charging stations. Further in our proposed solution, we handle the constraints (such as charging sequence of EVs) by integrating a Constraint Programming (CP) model in our heuristic approach.

The rest of the paper is organized as follows. In the next section, we discuss the related literature for solving EV routing. The succeeding sections present the problem formulation as an MIP considering the limited battery capacities of EVs and waiting times at the charging stations. Next, we present a GA method to solve large problem instances and the results gathered from the numerical experiments. The last sections concludes the paper with avenues of future research.

2   Related Literature

The Electric Vehicle Routing Problem (EVRP) is a special case of the traditional Vehicle Routing Problem with the additional set constraints due to the electric engine technology. Schneider et al. [1] describe the EVRP with Time

Windows (EVRPTW) as an MIP formulation and propose a heuristic combining variable neighborhood search and tabu search. In their problem formulation, the authors considered charging stations having no capacity constraint on number of chargers. In addition, they have not considered any waiting times at the charging stations. Goeke and Schneider [2] extended the EVRPTW model for a mixed fleet of conventional and electric vehicles with time windows and capacity constraint. The authors emphasize on the energy consumption models for the conventional and electric vehicles which is non-linear by nature. Sassi and Oulamara [3] considered an electric vehicle scheduling and optimal charging problem in order to assign vehicles to tours and minimize the charging cost. In their model, they considered constraints related to chargers, electricity grid, and EVs driving range.

In [4], the authors discuss the formulation of a mixed fleet with a case study at the city of Amsterdam. In their objective, the authors factors in the fixed cost of vehicles and the variable cost which depends on the en-route time and distance. Although the authors consider the battery capacity constraints of the electric vehicles, there is no consideration given to the amount of time a vehicle may need to spend at a charging station. The authors in [5] considers partial charging of electric vehicles with non-linear charging functions. The non-linearity of the battery charging function is handled using a piece-wise linear approximation. The authors model the problem to minimize the total time which includes the travel time and time spent at the charging stations.

Qin and Zhang [6] discuss the scheduling of charging activities with minimal waiting time of EVs in a network of EVs and charging stations. The authors consider EVs in general not involving a logistic fleet operation. Although, the authors address the problem of waiting time at the charging stations which translates to the availability of the charging outlets, the routing problem of the vehicle is not considered. Additionally, Froger et al. [7] address the problem of limited capacity charging stations for vehicle routing and provide matheuristic based approach to solve the problem. EVRP can be considered a specialized problem of the larger set of the Green Vehicle Routing Problem (G-VRP) that focuses on the use of alternative fuel sources for the vehicles. Hence, G-VRP also focuses on similar issues such as the limits to driving range and limited refueling stations. In [8] for example, the authors discuss the G-VRP and provide an MIP formulation for this problem.

As can be noted from the discussions above, the literature on EVs is segregated in two broad groups - those who consider it as a routing problem for fleet of EVs and those who consider the problem of scheduling and congestion at the charging stations. There is also a third variant which is related to the placement of charging stations infrastructure such as the one discussed in [9]. However, the problem of restricted number of charging outlets at the charging stations for the EV fleet operations does not seem to be addressed in the literature. The waiting times due to unavailability of charging outlets or scheduling the vehicle at the charging stations while planning their routes in a traditional VRP setting warrants further considerations to come up with real-world solutions.

Table 1. Variables and parameters of our proposed EVRP model

Notations	Description
$0, N+1$	Start and end depot
V	Set of customer nodes $\{1 \ldots N\}$
V_0	Set of customer nodes with start depot
V_{N+1}	Set of customer nodes with end depot
$V_{0,N+1}$	Set of customer nodes with start and end depots
V'	Set of nodes - the customer nodes and charging station outlets associated dummy nodes
V'_0	Set of nodes - the customer nodes, charging stations outlets associated dummy nodes, and depot $\{0\}$
V'_{N+1}	Set of nodes - the customer nodes, charging stations outlets associated dummy nodes, and depot $\{N+1\}$
Ω	Set of vehicles $\{1 \ldots W\}$
R	Set of charging station outlets $\{1 \ldots K\}$
R^k	Set of dummy nodes (indicate potential vehicles visit for a station) for charging station outlet k, set size equals to W, and $W \leq N$
Parameters	
T	Maximum time duration of a route
C	Cargo capacity of each vehicle
B	Battery capacity of each vehicle
δ	Rate of battery discharge
β	Rate of battery charging
s_i	Service time at node i
d_{ij}	Distance between two nodes i and j
t_{ij}	Travel time between two nodes i and j
D_i	Demand at node i
Variables	
x_{ij}	Binary decision variable for visit from node i to j; 1 if (i,j) is traveled, 0 otherwise
τ_i	Decision variable specifying the time of arrival at node i
$\tau^\hbar_{N+1}$	Decision variable specifying the time of arrival at the end depot for vehicle $\hbar$
w_i	Decision variable specifying the time of start of charging at the charging station outlet i
f_i	Decision variable specifying the remaining cargo on arrival at node i
y_i	Decision variable specifying the remaining charge available on arrival at node i
b^k_{uv}	Binary decision variable for a vehicle visiting the uth dummy node of charging station outlet k in sequence (order) index v; 1 if u takes the sequence (order) index v and 0 otherwise
z^k_u	Decision variable specifying the sequence (order) of charging service at station outlet k for each $u \in R^k$
$a^1_{m,u,v}, a^2_{m,u,v}, a_{m,u,v}$	Intermediate binary decision variables for deriving the service sequence of visits at a charging station outlet
a_{uv}	Binary decision variable to derive the sequence of visit at the charging station; 1 if dummy node v is visited after u and 0 otherwise

3   Mathematical Model

The primary objective of our work is to address the limited capacities of the charging stations which may result in the formation of queues in order to service a logistics fleet of EVs. A charging station consists of a set of charging outlets wherein a vehicle is plugged in to recharge its battery. Such charging outlets could be limited in number at the charging stations depending upon the electric grid capacity and charging stations' space constraints. In this section, we provide the exact mathematical model for our EVRP considering waiting times of the vehicles at the charging stations.

Let V be the set of customer nodes and R be the set of charging stations. Let the start and end depot node instances be denoted by 0 and $N + 1$. We will use 0 and $N + 1$ in the subscript to denote the set of nodes with start or end depot instance or both as in [1]. Table 1 lists the variables and parameters used in our model. To allow multiple visits to the charging station outlets, we create dummy nodes for every charging stations outlet in R. We further consider a linear discharge of the vehicle batteries and the distance among the nodes is Euclidean. We assume that the vehicles leave the depot with full charge and will not visit charging station outlet before completing their first delivery. Moreover, each visit to the charging station outlet results in full battery charge of the vehicles.

The objective function in our mixed integer program is to minimize the sum of route durations, which includes the travel time, the service time at the customer nodes, the charging and waiting times at the charging station outlets.

$$\min \sum_{\hbar \in \Omega} \tau_{N+1}^{\hbar} \tag{1}$$

Subject to:

$$\sum_{j \in V'_{N+1}} x_{ij} = 1, \forall i \in V \tag{2}$$

$$\sum_{j \in V'_{N+1}} x_{ij} \leq 1, \forall i \in R \tag{3}$$

$$\sum_{i \in V'_{N+1}} x_{ji} - \sum_{i \in V'_0} x_{ij} = 0, \forall j \in V' \tag{4}$$

$$\tau_i + (t_{ij} + s_i)x_{ij} - T(1 - x_{ij}) \leq \tau_j, \forall i \in V_0, \forall j \in V' \tag{5}$$

$$w_i + \beta(B - y_i) + t_{ij}x_{ij} - (T + \beta B)(1 - x_{ij}) \leq \tau_j, \forall i \in V_0, \forall j \in V'_{N+1} \tag{6}$$

$$\tau_i + (t_{i,N+1} + s_i)x_{i,N+1} - T(1 - x_{i,N+1}) \leq \tau_{N+1}^{\hbar}, \forall i \in V, \hbar \in \Omega \tag{7}$$

$$w_i + \beta(B - y_i) + t_{i,N+1}x_{i,N+1} - (T + \beta B)(1 - x_{i,N+1}) \leq \tau_{N+1}^{\hbar}, \forall i \in \bigcup_{k \in R} R^k, \hbar \in \Omega \tag{8}$$

$$\tau_i \leq w_i, \forall i \in R^k \tag{9}$$

$$u_i - D_i x_{ij} + C(1 - x_{ij}) \geq f_j \geq 0, \forall i \in V_0', \forall j \in V_{N+1}' \tag{10}$$

$$C \geq f_0 \geq 0 \tag{11}$$

$$y_i - \delta d_{ij} x_{ij} + B(1 - x_{ij}) \geq y_j \geq 0, \forall j \in V_{N+1}', \forall i \in V \tag{12}$$

$$B - \delta d_{ij} x_{ij} \geq y_j \geq 0, \forall j \in V_{N+1}', \forall i \in R^k, k \in R \tag{13}$$

$$\sum_{u \in R^k} b_{uv}^k \leq 1, \forall v \in \Omega, \forall k \in R \tag{14}$$

$$\sum_{v \in V} b_{uv}^k - \sum_{v \in V} x_{uv} = 0, \forall u \in R^k, \forall k \in R \tag{15}$$

$$\sum_{v \in V} v b_{uv}^k = z_u^k, \forall u \in R^k, \forall k \in R \tag{16}$$

$$z_v^k - z_u^k - (m - 1) - M a_{m,u,v}^1 \leq 0, \forall m \in \Omega \backslash N, u, v \in R^k \tag{17}$$

$$(m + 1) - (z_v^k - z_u^k) - M a_{m,u,v}^2 \leq 0, \forall m \in \Omega \backslash N, u, v \in R^k \tag{18}$$

$$a_{m,u,v}^1 + a_{m,u,v}^2 - a_{m,u,v} \leq 1, \forall m \in \Omega \backslash N, u, v \in R^k \tag{19}$$

$$\sum_{m \in \Omega \backslash N} a_{m,u,v} = a_{uv}, \forall u, v \in R^k \tag{20}$$

$$(a_{uv} - 1)M \leq \tau_v - \tau_u, \ \forall u, v \in R^k \tag{21}$$

$$w_v - w_u + (1 - a_{uv})M \geq (B - y_u)\beta, \forall u, v \in R^k \tag{22}$$

$$x_{ij} \in \{0,1\}, \forall i \in V_0', j \in V_{N+1}' \quad b_{uv}^k \in \{0,1\}, \forall u \in R^k, v \in \Omega, k \in R \tag{23}$$

Equation (1) minimizes the sum of route duration of each individual route. Constraint (2) specifies that all the customer nodes must be visited. Constraint (3) defines the visit to the charging station outlets. The constraint in (4) ensures flow conservation, which basically means all the outflows from a node and inflows to a node must be equal i.e. all the vehicles that leave a node must be equal to the number of vehicles that arrived at that node. Constraints (5) and (6) provide the arrival time at a node from a customer node and charging station outlet respectively. Constraints (7) and (8) provide the arrival time at the depot from a customer node and charging station outlet. Constraint (9) forces that the start of charging at the charging station outlet must be after the arrival and possible waiting. Constraints (10) and (11) is for the fulfillment of customer demand while constraints (12) and (13) are to ensure that the vehicles batteries always have positive charge.

Constraints (14)–(22) are meant to derive the waiting times of the vehicles that queue at a charging station outlet. The basic idea here is to sequence the vehicle based on their arrivals at the charging station outlet and use this sequence to calculate the start of charging of the vehicles at the charging station outlet. Constraint (14) forces that whenever a vehicle visits the charging station outlet, it gets assigned a sequence index. Constraint (15) ensures that all the vehicles that has arrived, also leave the charging station outlet, while (16) assigns the sequence index to the vehicles that arrive at the charging station outlet after serving a given customer node. Constraints (17)–(20) ensure the vehicle sequence of charging at the charging station outlet. Finally, constraints (21)–(22) calculate the time to start the charging based on the sequence index of the vehicles.

The binary variables $a^1_{m,u,v}$, $a^2_{m,u,v}$ and $a_{m,u,v}$ are introduced to get the visiting order of the EVs at a charging station outlet based on m. The associated constraints (17)–(20) ensures that, if $a^1_{m,u,v}$ equals to 1, $Z^k_v - Z^k_u > m - 1$, similarly, if $a^2_{m,u,v}$ equals to 1, it means $Z^k_v - Z^k_u < m + 1$, and if $a_{m,u,v}$ equals to 1, it means $Z^k_v - Z^k_u = m + 1$. Constraint (17–20) is to define the sequence of vehicles visit at charge station and deduct the waiting time. Note that m belongs to the set of $\{1,\ldots, N\text{-}1\}$, hence, the big M in (17) and (18) can be defined as N.

4 Genetic Algorithm (GA)

In this Section, we present our Genetic Algorithm (GA) approach to solve our EVRP problem defined above. The initial population size is set to Λ. We use a greedy algorithm to generate the initial population of solutions, which vary the order of selection of customer nodes to generate different individuals of the population. The fitness function used to evaluate an individual solution is the total route duration of all the routes to serve all the customers in the problem instance. The children solutions are improved by applying a local search heuristics and battery charge station visiting algorithm.

To generate an initial solution, we first assign an empty route that start and end at depot to each vehicle. We then randomly insert a customer node at the best position of the route for the vehicle with the smallest additional cost. If the battery charge of the vehicle cannot support visiting to the next customer, we add a charging station outlet visit. Once the route has been updated, we block the corresponding charging station outlet time interval occupied by this vehicle. When no more customer nodes could be added to the first vehicle, we proceed to the second vehicle, and so on. We repeat this process until all customers are fully served.

We represent each individual solution (chromosome) as a sequence of node labels. Hence, each chromosome starts with a depot label and ends with a depot label. All the customers and stations labels are inserted in between based on the routes taken. The end/start of each route is marked by the depot label. Hence, a single chromosome can be divided into a set of segments with starting and ending depot labels. Each such segment represents a vehicle route. Figure 1 depicts the representation of a chromosome with three routes, two charging station outlets and eight customer nodes wherein **D** denotes depot, **C** denotes customer nodes, and **S** denotes the charging station outlets. Hence, when we evaluate the fitness function, we will translate the chromosome to multi-routes, which is split by **D**, then, we calculate the sum of route durations.

D	C1	C2	S1	C3	S2	D	C5	S2	C4	C6	D	C7	S1	C8	D

Fig. 1. An individual solution representation in the GA

Table 2. Variables and parameters of the CP model

Parameters	
$succ_i$	Successive node in a path after serving customer i
$t_{i,succ_i}$	Travel time from node i to the successive node
λ_i	Battery charging time plus service time associated with node i
Variables	
arr_i	Earliest arrival time to start service at a node i
$serv_i$	Service start time at node i
s_k	Start depot node of vehicle k
e_k	End depot nodes of vehicle k
R	Set of charging station
R^o	Set of nodes (dummy nodes) of oth charging location
C^o	Charging station outlet capacity of oth charging location
V	Set of nodes in the vehicles route
E	Set of end depot nodes in the vehicles route

4.1   Constraint Programming (CP) for Fitness Evaluation

We vary the order of selection of customer nodes to generate different individuals of the population. The fitness function used to evaluate an individual solution is the total route duration to serve all the customers in the problem instance.

Let $c(s)$ be the routing cost (that is, sum of the travel time and service time) of route s. The fitness function f is defined to be the sum of routing costs and waiting times of all routes, each route has a fitness value $f(s) = c(s)+w(s)$, where $w(s)$ is the total waiting time incurred for route s. Interestingly, for a route s, $c(s)$ is easy to calculate, but the calculation of waiting time is complicated, as vehicles can visit charging station outlet many times, and wait either at charging station or customer locations. This makes the calculation of the consequent waiting times (if any) non-trivial. For this purpose, we apply a Constraint Programming (CP) model for calculating the optimal fitness value.

For the CP model, we first find that the solution satisfies the capacity and battery charge constraints for the vehicle. If the constraints are satisfied, then, we apply the CP model to calculate the fitness value of the total route duration.

$$\min \sum_{k \in \Omega} serv_{e_k} - arr_{s_k} \qquad (24)$$

Subject to:

$$arr_i \leq serv_i, \forall\, i \in V \tag{25}$$
$$serv_i + \lambda_i + t_{i,succ_i} = arr_{succ_i}, \forall\, i \in V \backslash E \tag{26}$$
$$CUMULATIVE(\{serv_i, \lambda_i : i \in R^o\}, 1, C^o), \forall\, R^o \in R \tag{27}$$

The variables and parameters used in the CP model are listed in Table 2. The objective function (24) minimizes the total route duration. Constraints (25)–(26) define the earliest arrival time, and start time of charging at node i. The charging resource constraint (27) is modeled using the global constraint, where $serv_i$, λ_i, 1 and C^o represent the start time of charging, duration, resource requirement, and the capacity of the resource respectively. The cumulative constraint (27) specifies the requirements on tasks which need to be scheduled on a number of resources. It expresses the fact that at any time instant the total use of these resources for the tasks does not exceed a given limit. For our problem, we have a list of charging tasks, each task requires only one resource, and the upper limit of the amount of charging resources limit equals to C^o.

The average running time for our CP model is around 2 seconds for 100-nodes problem instances. Hence, we provide the following model simplification procedure before running the CP model: (1) If a vehicle never visits any charging station outlet, we can remove the whole path of vehicle k from the model; (2) If there are fewer than C^o visits for o^{th} charging station, it is unnecessary to check the cumulative constraint (27) as the constraint is automatically satisfied; (3) Suppose a charging station has only been visited by the same vehicle k, then we can also remove the whole path of vehicle k since there is no dependency on other vehicles.

Algorithm 1. Virtual charging station outlet booking algorithm

---

Input: battery charge task i, earliest service start time τ_i, service duration λ_i,
      initial $\varpi_i = 0$, $\overline{v}_i = 0$

1   if charge station ϵ_j is available at time interval $[\tau_i, \tau_i + \lambda_i]$
2      $\varpi_i = 0$;
3   assign battery charge task i to time interval $[\tau_i, \tau_i + \lambda_i]$ on ϵ_j;
4   **stop**
5   **end if**
6 Let Θ be the set of booked battery charge tasks that intersect with time τ_i, and
     τ_i' denotes the earliest completion time of tasks in Θ
7 if $[\tau_i', \tau_i' + \lambda_i]$ is within the planning horizon (no penalty)
8   assign battery charge task i to $[\tau_i', \tau_i' + \lambda_i]$
9   record waiting time ϖ_i, and update charging outlet booking
10 **else**
11   assign battery charge task i after τ_i
12 record waiting time ϖ_i and penalty $\overline{v}_i$, and update charging outlet booking
13 **end if**
    Output: ϖ_i and $\overline{v}_i$

Unfortunately, it is still quite time-consuming to apply CP to compute the exact waiting time in every iteration, hence, we only trigger CP under two cases: (1) when we manage to find a solution that satisfies vehicle capacity and time window constraints; and (2) after every p (a parameter that we set default value to 20) iterations. In other cases, we apply a heuristic to estimate the waiting times. The basic idea of this heuristic is to simulate a virtual booking system that will compute an approximated waiting time and station capacity violation whenever a vehicle visits a charging station as follows.

Heuristic for Charging Station Waiting Time Calculation. Let us define a charging station visit as a special task. Since waiting time is a global variable that depends on the arrival times at *all* charging station visited by all vehicles, we are not able to compute the exact waiting time by simply considering each trip independently. Therefore, for a given charging task i in the solution, we apply Algorithm 1 to estimate the waiting time. Let τ_i be the earliest service start time for charging task i, and λ_i be the charge time at charge station m.

- Case 1: charging station m has an available (i.e. unoccupied) time interval at $[\tau_i, \tau_i + \lambda_i]$.
  This is very straightforward - we assign $[\tau_i, \tau_i + \lambda_i]$ to charging task i (shown in lines 1–5 in Algorithm 1).
- Case 2: charging station m has an available time interval with length at least λ_i but later than τ_i. Let Θ be the set of booked charging outlets intersect with time τ_i, and τ_i' denote the earliest completion time of charging tasks in Θ. We assign $[\tau_i', \tau_i' + \lambda_i]$ to charging task i (shown in lines 6–8).
- Case 3: there is no available time interval that can fit charging task i.
  We assign task i after time τ_i (line 11). Note that by doing so, the booked charging outlets time interval will be overlapping (i.e. the solution is infeasible). Therefore, penalties need to be imposed (line 12).

Overall, the total waiting time $w(s)$ is evaluated by sum of all tasks ϖ_i and $\overline{v_i}$.

4.2   Selection and Crossover

The top K elite individuals are kept for the next generation based on their fitness. All the individuals will participate to produce new individuals generated by applying the crossover operator. Noted that the crossover rate is 100%, as we will apply our local search operator in Sect. 4.4, which is more powerful than general mutation operator in the GA. The pair of parents for crossover are chosen using tournament selection method. This means, we randomly sample two individuals from the individuals. The partially matched crossover operator is used to create two new individuals to replace the parents. For given parents P_1 and P_2, two cutting points are chosen. With the middle sub-path kept intact, we swap the arc and nodes between two parents. Figure 2 shows an example. The top two rows represent parents individuals, and bottom two rows stand for the children individuals. The crossover operation results in the creation of new children. All the customers must be served in our problem instance, hence if left arcs or nodes cannot be inserted to the child because of vehicle load or vehicle available time infeasibility, we add empty vehicles and insert the leftover arcs/nodes to the empty vehicles. The two new individuals along with the current population form the new population for the next generation of GA. We terminate GA when there is no improvement in the best solution for a fixed number of generations or the number of iterations reaches the upper limit.

4.3   Columns Based Chromosome Generation

In every 1000 iterations of GA generations, we generate new chromosomes by recording the last 1000 iterations of feasible vehicle paths in the population as columns. We find the best combination of those columns, generate new chromosomes and add them to the population. Let R be the set of all columns, V be the set of customers included in the columns, and P stands for presetting price of a customer. Parameters include c_r and a_r^i, where c_r equals to the travel time of column and r, a_r^i is a parameter that equals 0 or 1, where 1 indicates node i is served by column r. Binary decision variables include y_i and z_r, y_i which equals to 1 if customer i is selected. Similarly, z_r equals to 1 if column (trip) r is selected. We aim to generate better solutions from a population of historical solutions, with the Set Covering formulation shown as follows:

$$\max \sum_{i \in V} P y_i - \sum_{r \in R} c_r z_r \tag{28}$$

Subject to:

$$\sum_{r \in R} a_r^i z_r = y_i, \quad \forall\, i \in V \tag{29}$$

We assume serving a customer leads to revenue P, hence the objective function (28) maximize the revenue minus the total travel time. We use y_i to check customer i be served or not in Constraints (29). If not all the customers can be served, we will insert the leftover customers to the solution one by one to the position with the smallest insertion cost added.

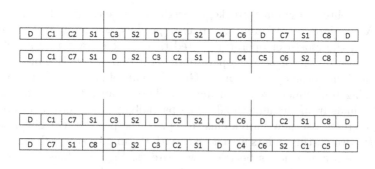

Fig. 2. Crossover operator

4.4   Local Search on Solutions

Once a new chromosome been generated, we improve the solution by local search. Customer nodes are selected and added to a perturbation set. To select the customer nodes, we use an objective value based operator. By checking the vehicle routes, we calculate the objective value after removing one customer node, then, descending sort the values, we choose the top 10% customer nodes and added to the perturbation set.

After the customer nodes selection, three perturbation operators (I1-I3) are used. The probability of using first operator is set to 0.5 while the probability of using second and third operators is set to 0.25. We accept the new solution if the fitness value is better after applying local search.

1-by-1 (I1): The selected customer nodes are sequentially removed one by one and reinserted into the best position (the highest improvement for the current objective value).

1-by-1 different vehicle (I2): Suppose, customer node i is removed from some route k, this operator tries to insert the customer node i into different route (which is not k) in a better position.

1-by-1 same vehicle (I3): Suppose, customer node i is removed from some route k, this operator tries to insert the node i into the same route k again but in a better position.

4.5   Management of Charging Station Visits

If no feasible solution can be obtained because of the violation of battery limit constraint, we add charging station outlet visits (described in Algorithm 2) to fix the infeasibility. Basically, we include charging station visits if only the vehicles cannot reach to the customer nodes ρ_i with the available charge in their batteries. The position to insert the charging station visit can be any location before ρ_i, hence, we find the best position to insert the closest charging station (line 4–9 of Algorithm 2). In case, a charging station cannot be inserted (due to long distance) then we split the visits to customers to two or more vehicles.

During the GA iterations, once a charging station has been removed, we reinsert it only if the same would lead to a better fitness value. Besides that, every 20 iterations, we double check the current iteration solution, and remove useless charging stations and insert new charging station. Once a new best solution or new best feasible solution has been found, the charging station visit gets updated as well.

Algorithm 2. Insert charge station visits in a given vehicle path

---

Input: path $(\rho_1, \rho_2, \ldots, \rho_N)$, fitness value $f(s)$ and $f(s)_{old}$ (both equals to fitness value of path s)

1 **do**
2   **for all** $\rho_1, \rho_2, \ldots, \rho_N$ **do**
3   check the battery when vehicle arrive at location ρ_i
4   **if** battery is lower than 0
5   find the best charge station in $1, \ldots, K$ and best insertion point between $\rho_1, \ldots, \rho_{i-1}$ with smallest fitness value
6   **if** lowest fitness value is smaller than $f(s)_{old}$
7   insert charge station, update path s, and $f(s)_{old} := f(s)$, $f(s) :=$ smallest fitness value
8   **else** break;
9   **end if**
10   **end for**
11 **while** $f(s) < f(s)_{old}$
Output: $f(s)$ and updated path s

---

5 Numerical Experiments

To evaluate our approach, we performed numerical experiments using two sets of problem instances on an Intel Xeon E5-2667v4 8C/16T (3.2 GHz) 16 core CPU 32 GB RAM machine. We used a batch file to utilize all the 16 cores while running the numerical experiments wherein each instance used one single core. The first set is the self-generated small instances (having 5 and 10 number of customers), aimed to compare the solutions between the heuristic approach and math model (using CPLEX solver version 12.8.0) along with the verification of the solution details such as travel time, waiting time and charging time. The second set is based on instances provided by Schneider et al. [1] that are adapted from Solomon [10] instances to include charging stations. These instances are provided in three different categories with 100 customer nodes. With our specific problem setup that focuses on limited charging capacity at the charging stations, we realized that generating instances suitable this problem is in itself a complex problem. As the literature do not provide Solomon instances with more than 100 customer nodes, we only focus on the variety of 100 nodes instances for our experimentation. According to our problem definition, we ignore the customer time windows available in the Solomon instances. All test instances can be found at https://unicen.smu.edu.sg/research/urban-logistics/electric-vehicles-routing-problem-waiting-times-charging-stations.

For the first set of instances, the battery capacity was fixed at 30, charging station outlet capacity equals to 1, and the vehicle capacity was set to 100. The service time at all the customer nodes was set to 10 time units. The second set of instances used the parameters as defined in the type "C", "R" and "RC" of Solomon instances. The vehicle battery capacity was set to 62.14 or 79.69, number of charging station locations was set to a random number between 5–18, charging station outlet capacity equals to 1, the location is randomly chosen within the spatial distribution, and charging rate was set to full charge battery divided by average customer service time. For GA, we use the following parameter values: initial population size $\Lambda = 100$, number of top elite individual $K = 20$, number of iterations was set to 5000, penalty parameter equals to 20 for waiting time approximation, and customer price in column based operator was set to twice of average travel time between nodes. All the experiments were performed for 10 times and the best result from the multiple runs are reported. Before discussing the results, we depict the evolution of GA in Fig. 3 for one typical instance. As can be seen, the fitness value decreases in the number of generations, and improves from 5979 to 2218 in about 3000 generations and stop improving afterwards.

5.1 Results

Table 3 compares the results from the exact math model solved using Cplex and the GA based heuristics for the first set of small instances. The table shows the fitness value which is the sum of route durations, travel time (TT), service time (ST), the combined waiting and charging time (WCT), and the running time in

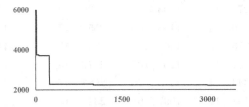

Fig. 3. GA converge curve

Table 3. Cplex and GA heuristic based solution on 5 and 10 nodes instances

Instance	Cplex					GA				
	Vehicles	Fitness	TT	WCT	Time (sec)	Vehicles	Fitness	TT	WCT	Time (sec)
5–1	3	405	196	159	165	4	427	228	149	10
5–2	3	400	195	155	315	3	400	180	170	11
5–3	4	522	236	236	46	4	554	229	275	16
5–4	4	493	245	198	44	4	497	227	220	11
5–5	4	489	246	193	22	4	489	223	216	19
10–1	4	747	358	289	10800	5	696	316	330	24
10–2	4	783	365	328	10800	4	702	332	270	18
10–3	4	1160	546	514	10800	4	571	280	191	32
10–4	4	984	437	450	10800	4	943	354	489	29
10–5	3	665	318	247	10800	3	622	286	246	35

Table 4. Results of GA based heuristics solution on the 100-node Solomon instances.

Instance	Best Obj.	No. Veh.	TT	WCT	Ave. Obj.	Ave. No. Veh.	Ave. TT	Ave. WCT
c101	10753	12	1133	620	11234	15	1373	862
c102	10475	13	1200	275	10974	16	1540	434
c103	10740	12	1140	600	11712	22	1895	817
c104	10732	15	1255	477	11709	22	1906	803
c105	10698	13	1328	470	11104	13	1390	714
c106	10548	11	1085	463	11064	13	1477	588
c107	10883	17	1497	386	11684	22	1757	927
c108	10654	13	1123	531	11317	25	1902	416
c109	10843	19	1244	599	11319	18	1657	662
r101	2329	14	1246	83	2941	18	1765	176
r102	2385	16	1320	65	2700	18	1587	113
r103	2467	13	1353	114	2885	20	1542	343
r104	2315	14	1216	99	2431	15	1345	86
r105	2392	17	1322	70	2931	22	1789	142
r106	2284	12	1192	92	2605	17	1509	96
r107	2270	14	1193	77	2589	17	1475	115
r108	2250	14	1185	65	2559	15	1457	102
r109	2375	15	1263	112	2545	15	1442	103
r110	2312	14	1209	103	2572	18	1493	79
r111	2343	14	1273	70	2683	17	1561	122
r112	2158	12	1097	61	2324	14	1239	86
rc101	2667	15	1608	59	2883	16	1787	96
rc102	2748	17	1655	93	3038	18	1897	141
rc103	2729	18	1657	72	3088	17	1968	120
rc104	2730	15	1645	85	2902	16	1805	98
rc105	2847	17	1746	101	3175	19	2015	160
rc106	2997	18	1843	154	3199	19	2041	158
rc107	2820	15	1726	94	2960	16	1838	122
rc108	2768	15	1690	78	3207	19	2083	124

Ave.: Average, Obj.: Objective value, TT: travel time
WCT: Combined waiting and charging time, No. Veh.: Number of Vehicles

seconds for all different instances. As can be observed the gap between the travel time for Cplex and GA is quite comparable for the instances with 5 customer nodes. A similar trend can be noted for the waiting and charging time. However, for instances with 10 customer nodes, GA based solutions seems to spend less time in travel as well as waiting and charging. The results shows that considering waiting time at the charging stations while planning the routes for EV based fleet is crucial as the amount of time spend in waiting and charging is almost equal to the travel time.

One can see, for the 5 customer node instances, Cplex can solve it quickly, while for the instances with 10 customer nodes, we have to stop Cplex after 3 hrs (10800 secs) and get the best solution discovered so far. In comparison to the best solution found by Cplex, the GA heuristics fairs well with very less execution time. For problem instances with 5 customer nodes, there are very low gap between the solutions while in case of instances with 10 customer nodes, GA always finds a better quality solution in very less time as compared to the Cplex.

Table 4 lists the best and average results for the set of Solomon instances using the GA based heuristics approach. The best objective value (from 10 runs) is the sum of the route durations of all the vehicles after they return to the depot. On average, the CPU time equals to 10 min overall instances. Since our objective value involve charging and waiting times at the charging stations, the same is not directly comparable with the literature [10]. We observed that the "R" group Solomon instances have longest travel time, while "C" and "RC" group instances have similar travel times. This happens as the distribution of "R" group Solomon instances are more disperse than other two groups. Further, the waiting times of "C" group instances are much longer than other two groups, mainly because of the difference in the rate of charging.

6 Conclusions

The advancement of electric vehicles technology has paved the way forward for EV based logistic fleets for customer delivery. However, the route planning with EVs brings forward new challenges to the already complex problem of vehicle routing. In this paper, we consider the problem of formation of vehicle queues at the charging stations en-route to customers. We consider the extra waiting time due to the queue at the charging stations to formulate an exact mathematical model for route planning of EV based logistics fleet. The evaluations are performed on a set of problem instances catering to specific requirement of our problem setup. Although it is difficult to get a solution from the exact mathematical model beyond 10 customer nodes, the GA heuristic based solution has shown comparable performance and can be quickly executed for large size instances. In the future, we plan to extend our work by adding more practical constraints in our model such as customer time-windows, partial charging of vehicle at the charging stations, etc. Further, using the current results as benchmark, more comparative evaluations with variety of instances can be carried out in order to validate the scalability and efficiency of our GA based heuristics approach.

Acknowledgement. This research is funded by the National Research Foundation Singapore under its Corp Lab @ University scheme and Fujitsu Limited as part of the A*STAR-Fujitsu-SMU Urban Computing and Engineering Centre of Excellence.

References

1. Schneider, M., Stenger, A., Goeke, D.: The electric vehicle-routing problem with time windows and recharging stations. Transp. Sci. **48**(4), 500–520 (2014)
2. Goeke, D., Schneider, M.: Routing a mixed fleet of electric and conventional vehicles. Eur. J. Oper. Res. **245**(1), 81–99 (2015)
3. Sassi, O., Oulamara, A.: Electric vehicle scheduling and optimal charging problem: complexity, exact and heuristic approaches. Int. J. Prod. Res. **55**(2), 519–535 (2017)
4. Van Duin, J., Tavasszy, L.A., Quak, H.: Towards electric-urban freight: first promising steps in the electric vehicle revolution (2013)
5. Montoya, A., Guéret, C., Mendoza, J., Villegas, J.: The electric vehicle routing problem with partial charging and nonlinear charging function. Ph.D. thesis, LARIS (2015)
6. Qin, H., Zhang, W.: Charging scheduling with minimal waiting in a network of electric vehicles and charging stations. In: Proceedings of the Eighth ACM international workshop on Vehicular inter-networking, pp. 51–60. ACM (2011)
7. Froger, A., Mendoza, J.E., Jabali, O., Laporte, G.: A matheuristic for the electric vehicle routing problem with capacitated charging stations. Research report, Centre interuniversitaire de recherche sur les reseaux d'entreprise, la logistique et le transport (CIRRELT), June 2017. https://hal.archives-ouvertes.fr/hal-01559524
8. Erdoğan, S., Miller-Hooks, E.: A green vehicle routing problem. Transp. Res. Part E Logist. Transp. Rev. **48**(1), 100–114 (2012)
9. Xiong, Y., Gan, J., An, B., Miao, C., Bazzan, A.L.: Optimal electric vehicle charging station placement. In: IJCAI, pp. 2662–2668 (2015)
10. Solomon benchmark 100 nodes. https://www.sintef.no/projectweb/top/vrptw/solomon-benchmark/100-customers/. Accessed 15 Jun 2018

Multiple Periods Vehicle Routing Problems: A Case Study

Bilal Messaoudi[1,2]([⊠]) [ID], Ammar Oulamara[1], and Nastaran Rahmani[2]

[1] Universiy of Lorraine - LORIA UMR 7503, Campus Scientifique,
615 Rue du Jardin-Botanique, 54506 Vandoeuvre-les-Nancy, France
`bilal.messaoudi@univ-lorraine.fr`
[2] Antsway SA, ARTEM - 92, Rue du Sergent Blandan, 54042 Nancy Cedex, France

Abstract. In this paper, we consider a challenging problem faced by a hygiene services company. The problem consists of planning and routing a set of customers over a 3-month horizon period where multiple frequencies of visits can be required simultaneously by each single customer. The objective is then threefold: (1) balancing workload between vehicles (agents) (2) minimizing number of visits to the same customer (3) minimizing total routing costs. In this context, a routing plan must be prepared for the whole horizon, taking into account all constraints of the problem. We model the problem using a decomposition approach of planning horizon, namely, weeks planning and days planning optimization. We propose an adaptive large neighborhood search with several operators for routing phase of solving approach. To evaluate the performance of the solving approach we solve an industrial instance with more than 6000 customers and 69951 requests of visits. The results show an excellent performance of the solving approach in terms of solution quality compared with the existing plan used by the hygiene services company.

Keywords: Vehicle routing problem · Planning and routing ·
Adaptive large neighborhood search

1 Introduction

Vehicle Routing Problems (VRP) have been widely considered in the literature due to their economic impact on real and industrial applications where optimized transport planning can save up to 20% of logistics costs. In some applications, visiting customers several times during a given planning horizon (one or several weeks) is necessary to overcome the reduction in storage space at the customer's site, or industrial client using a just-in-time production system. Furthermore, routing over a planning horizon allows smoothing customer demands and rational use of resources (vehicles, agents, etc.) without asking for external resources that generate additional costs. In addition, these routing problems enable grouping visits of

---

The original version of this chapter was revised: Errors in Section 4.1 and Equations 4 and 5 were corrected. The correction to this chapter is available at https://doi.org/10.1007/978-3-030-16711-0_14

© Springer Nature Switzerland AG 2019
A. Liefooghe and L. Paquete (Eds.): EvoCOP 2019, LNCS 11452, pp. 83–98, 2019.
https://doi.org/10.1007/978-3-030-16711-0_6

clients positioned close to each other and reducing routing costs of the fleet of vehi-cles. Considering several periods (days or weeks) of planning horizon makes routing problems more complex and introduces a new category of vehicle routing problem in the literature known as periodic or multiple periods vehicle routing problems.

In this paper, we study a routing problem of a French enterprise special-ized in the hygiene service for companies such as sanitary cleaning. Professional clients are visited with different frequencies depending on requested services, and planning of the routes occurs over a planning horizon of several days. Clients locations, together with their requested products and frequency of visits are known. A finite set of heterogeneous capacitated vehicles are available over the planning horizon, to serve clients. A solution to this problem consists of sets of routes that satisfy the demand and the frequency constraints, the capacity and the time shift of vehicles. The objective is to minimize the total costs of routes, while balancing workload over the planning horizon.

Clearly, this industrial case is a multilevel combinatorial optimization prob-lem. We propose a heuristic solution method that builds a solution in three phases, each of which concerns one of the decisions to be made. In the first phase, we decide on which week each client will be served, at the second phase, the assignment of clients to each day is decided over all weeks, and at the third phase a classical VRP for each day of the planning horizon is solved. The pro-posed heuristic solution method is numerically shown to be efficient on the indus-trial instance and an efficient improvement of the existing solution is proposed.

The rest of the paper is organized as follows: in Sect. 2, we briefly review the main works on periodic vehicle routing problems; in Sect. 3, we present the indus-trial case problem and provide a formal description and notations; in Sect. 4, the solution method is presented, which consists of our decomposition approach. Experimental results are presented in Sect. 5 and compared with the solution exploited by the company. Finally, some conclusions are drawn in Sect. 6.

2   Literature Review

In the periodic vehicle routing problem (PVRP), clients require visits on one or more days over a planning horizon of several days. Besides, there exists a set of feasible visit options for each client. First, each client must be assigned to a feasible visit option; a vehicle routing problem must then be solved for each day of the planning horizon, while total cost over the planning horizon is minimized. The periodic vehicle routing problem has received much attention in the literature, since it is used in many real-world applications such as waste collection [1], residential garbage collection [2], animal waste [3], replenish of retail stock [4], home healthcare nurses [5], maintenance of elevators [6]. More real application examples can be found in [7].

The periodic vehicle routing problem was first introduced by Beltrami and Bodin [8]. They considered one week planning horizon with the objective of minimizing both the number of used vehicles and the total travel time. They solved independent VRPs for each day of the period. Russell and Igo [9] gave a formal definition of the problem and proposed three heuristics that minimize the total distance travelled per week and balance vehicle requirements. Christofides and Beasley [10] presented an exact integer programming formulation of the periodic vehicle routing, and proposed a heuristic consisting of two steps to solve the problem. Russell and Gribbin [11] presented a solution method that consists of an initial route design, followed by three different improvement phases.

Meta-heuristic approaches are also developed for the PVRPs. Cordeau et al. [12] presented a tabu search method for solving the multi-depot vehicle routing problem (MDVRP) and the periodic traveling salesman problem (PTSP). Two types of neighborhood operators are proposed. Cordeau et al. [13], extended the work of [12] for PVRP with time windows. Francis et al. [14] developed an exact solution method for the PVRP with service choice. The solution method is based on Lagrangian relaxation and a branch-and-bound procedure is used to close the gap between the upper and the lower bounds. Alegre et al. [15] considered a periodic pick-up of raw materials for a manufacturer of automobile parts. They proposed a Scatter Search to solve this problem. The algorithm is based on a two-phase approach, the first phase assigns orders to days and the second phase constructs routes for each day. Mourgaya and Vanderbeck [16] developed a model to simultaneously address two objective criteria: balance of the workload across vehicles, regional compactness of the routes. The authors used a Dantzig-Wolfe reformulation and a column generation approach to solve the relaxed problem, and the resulting solution is rounded to produce a feasible solution to the PVRP by heuristically exploring the branch-and-bound tree. Hemmelmayr et al. [17] proposed a variable neighborhood search (VNS) algorithm using three neighborhood structures and the solution acceptance is based on simulated annealing. Vidal et al. [18] developed a hybrid genetic algorithm (GA) for the PVRP. A survey on periodic vehicle routing problems are proposed by Campbell and Wilson [7].

3   Problem Description and Notation

3.1   Industrial Problem

This paper focuses on a challenging problem encountered by a company that provides hygiene services to a set of professional clients. The company offers a sanitary cleaning service in addition to the distribution of products such as cleaning gel, toilet papers, in its clients' sanitary blocks, etc.

To solve the planning and routing problem, the company used software that did not match all of its operational constraints, so several manual changes were made on the solution to meet all constraints. Another disadvantage faced by the company was the quality of the solution in terms of total distance travelled, number of vehicles used, number of visits to each customer and balance of workload between drivers.

The company asked us to examine possible refinements of the company's service planning system in view of making it more cost effective, in order to provide improved customer service at a lower cost than that currently being experienced. The goal of this study is developing a new framework of optimization methods that include all operational constraints in route planning optimization that decrease manual changes in the planning after the optimization and route planning production, while the workload between drivers is balanced and total traveled distance, size of the fleet of used vehicles together with number of visits to each client are minimized.

The company serves 6000 clients over 12 weeks planning horizon; clients are geographically dispersed. Depending on sanitary blocks characteristics, environment, and usage, agents perform various cleaning tasks, in addition to replenish of cleaning gel and toilet papers. Company assigns each client to the same agent. Among the factors considered in assigning tasks to agents are the geographic proximity, agent workload and customers relations.

Each client requests a set of products. There are 14 different products, and the request of each client to each product is characterized by a quantity and a frequency over the planning horizon. Delivery frequency of the products takes value in the set $\{1, 2, 3, 6, 12\}$, where the frequency represents the number of visits over 12 weeks. The service duration of visits to clients consists of cleaning service time and delivery service time. Delivery service time depends on the type (frequency) and quantity of products to be delivered for each client, while cleaning service time is the time spend at each client independent of the type and quantity of the product. The latter is calculated only once per visit.

The company uses a fleet of 40 heterogeneous vehicles with different capacities. Each vehicle starts and finishes its route at the depot. The goal is to serve the demand of clients using available vehicles, respecting the following operational constraints:

1. *Vehicle capacity constraint*: states that the total demand of the customers on any route should not exceed the vehicle capacity.
2. *Route duration constraint*: ensures that the total duration of a route does not exceed an agent working shift duration.
3. *Same day visit constraint*: requires that each client must be visited on the same day of the week for all weeks when a visit is scheduled.
4. *One day visit constraint*: states that each client must be visited at most once per week.
5. *Same agent constraint*: assures that each customer is visited by the same agent for all weeks when a visit is scheduled.

3.2   Formal Description

A formal description of the problem can be defined as follows. Let $G = (V; E)$ be a network with $V = \{0, \ldots, n\}$ a set of $n + 1$ vertices where vertex 0 is the depot and vertices $1, \ldots, n$ represent clients to be visited, and E set of arcs, where each arc (i, j) represents the direct-travel possibility from i to j with travel time equal to t_{ij} and travel distance equal to d_{ij}. The planning period has a length of H weeks and each week has length of D days. A client is visited during this time period according to frequencies that depend on the type and quantities of requested products. At the depot, a set of m heterogeneous vehicles are available for servicing clients, and each vehicle v has a capacity c_v, $v = 1, \ldots, m$ and a maximum service duration T, where the service duration of the vehicle is computed as the sum of total travel time and total service times of visited clients. A visit of a client consists in delivering a set of products in addition to an on-site activity such as toilet cleaning. Furthermore, each client i is characterized by a time window for visits, a list P_i of requested products, and for each product $j \in P_i$, a frequency f_{ij} and quantity q_{ij} requested over the time horizon. The frequency f_{ij} defines the number of visits to clients i over the planning horizon H, for instance, if $f_{ij} = 2$, client i is delivered twice for product j during the time horizon with equal gaps between the visits. In our industrial case, the set of frequencies is $\{1, 2, 3, 6, 12\}$ over the time horizon of 12 weeks. Additional operational constraints are imposed by the industrial case study, covering agent (driver) working regulation such as working time, together with allowing visits of the same client every time by the same agent. The goal is to find a scenario of visits for every client and a set of routes for each day of the planning horizon, such that vehicle-capacity, route-duration and operational constraints are respected, where the number of visits to the same client, due to the multiple frequencies is minimized, workload between agents is balanced, the number of used vehicles as well as the total travel time is minimized.

4 Solution Approach

Due to the complex nature of the problem, which includes tactical and operational levels, solving it in one step, for example using a MIP solver, is impracticable, due to the complexity of constraints and the size of the actual instances to be solved which exceeds 6000 visits. We propose a decomposition approach to solve the problem, precisely, we develop a two-stage solution approach: in stage one clients are assigned to weeks depending on the requested products and frequency of each product, the clients of each week are then assigned to days. The second stage will solve the routing problem for each day of the planning horizon.

4.1 Weeks Planning Model

In this section, we focus on the problem of assigning clients to weeks with the objective of balancing workload over H weeks $(w_1, \ldots, w_H)$ of planning horizon while satisfying requests of customers. Given a set V of clients, and a set P of products, each client i asks for a subset of products P_i and for each product $p_{ij} \in P_i$ is associated a requested quantity q_{ij} and a delivery frequency f_{ij} over the H weeks of the planning horizon, where H is the lowest common multiple of $f_{ij}, \forall i \in V, \forall j \in P$. For each product p_{ij} with frequency f_{ij}, we define a set of possible delivery scenarios R_{ij}, where each scenario is a list of f_{ij} weeks, uniformly distributed over the planning horizon, and one scenario of R_{ij} is selected to supply product j to client i. For instance, if $f_{ij} = 2$, then the set of possible delivery scenarios of product p_{ij} is

$\{< w_1, w_{\frac{H}{2}+1} >, < w_2, w_{\frac{H}{2}+2} >, \ldots, < w_{\frac{H}{2}}, w_H >\}$ of weeks. In addition, selecting a scenario for a product p_{ij} is equivalent to selecting the first week w_{l_1}, and the kth visit occurs in week $w_{l_k} = w_{l_1} + k\frac{H}{f_{ij}}$, $k = 1, \ldots, f_{ij} - 1$. Thus, a solution of the weeks planning problem consists in selecting, for each product p_{ij} with frequency f_{ij}, the first visit in a set $\{1, 2, \ldots \frac{H}{f_{ij}}\}$ that minimizes the total number of clients' visits over the planning horizon, while respecting the capacity load of each week and balancing the workload over the planning horizon. We define $r_{ij} = \frac{H}{f_{ij}}$, $\forall i \in V$, $j \in P_i$ and for each client i we denote by g the index of the product with the highest frequency f_{ij}, i.e. $f_{ig} = \max\{f_{ij} \mid j \in P_i\}$. Let h_{ij}^g be the greatest common divisor between r_{ig} and r_{ij}, $\forall j \in P_i$ and $\forall i \in V$. Finally, we introduce the notation $u \equiv v[s]$ which means that $(u \mod s) = (v \mod s)$.

In the following, we use a MILP to model the weeks planning problem. The binary $x_{i,j,l}$ variable indicates whether or not the first delivery of the product j of client i is done on week l. The objective is minimizing the maximal service time of all clients over the weeks.

$$\min \quad L$$

$$\sum_{i \in V} \sum_{j \in P_i} (s_{ij} \cdot x_{i,j,l}) \le L \qquad\qquad l = 1, \ldots, H \qquad (1)$$

$$\sum_{i \in V} \sum_{j \in P_i} (q_{ij} \cdot x_{i,j,l}) \le C \qquad\qquad l = 1, \ldots, H \qquad (2)$$

$$\sum_{l=1}^{r_{ij}} x_{i,j,l} = 1 \qquad\qquad \forall i \in V, \forall j \in P_i \qquad (3)$$

$$x_{i,j,l} = x_{i,j,(l+k.r_{ij})} \qquad\qquad \begin{array}{c} \forall i \in V, \forall j \in P_i, l = 1, \ldots, r_{ij}, \\ \forall k \in \{1, \ldots, f_{ij} - 1\} \end{array} \qquad (4)$$

$$x_{i,g,l} \le \sum_{l' \in H_l} x_{i,j,l'} \qquad\qquad \begin{array}{c} \forall i \in V, \forall j \in P_i, l = 1, \ldots, r_{ig}, \\ H_l' = \{k = 1, \ldots, r_{ij} : k \equiv l[h_{ij}^g]\} \end{array} \qquad (5)$$

The first constraints (1) restrict the total workload of each week. Constraints (2) ensure that vehicles weekly capacity is not exceeded. Constraints (3) ensure that exactly one week is chosen for the first visit for each product and each customer and so that the customer is visited according to the frequency of the product in following weeks as stated in constraints (4). Constraints (5) restrict the number of visits to each client, depending on requested products and their frequencies. For instance, for a given client who requests two products p_1 and p_2 with frequencies 6 and 3, respectively, over 12 weeks planning period. The first visit of the client for product p_1 is either week 1 or week 2, while for product p_2, the first visit takes a value in the set $\{1, 2, 3, 4\}$. Thus, if the MILP chooses week 1 for p_1 then, the first visit of product p_2 is restricted to the set $\{1, 3\}$ so that the customer visit coincides with frequencies of the two products p_1 and p_2.

The weeks planning model is a generalization of the Bin Packing problem which is an NP-Hard problem. We can solve this MILP optimally using an optimization solver like CPLEX. Notice that the location of the clients is ignored in this model, however, this disadvantage is mitigated by the size of industrial instances (about 6000 visits) and low spatial spread of clients.

4.2   Days Planning Model

Given a weeks planning solution obtained in Sect. 4.1, in this section, the days planning model is presented, which consists in assigning clients to each day of the weeks. Notice that a client is assigned to the same day of the week over all the weeks of the planning horizon once client visits are scheduled (Sect. 4.1). The objective of days planning model is to balance the workload between days of the week.

Let C_w be the set of clients assigned to week w. Besides, S_{iw} and Q_{iw} denote the service time and the demand quantity of client i during the week w, respectively, obtained by solving the weeks planning model. Let C_{wf} be the set of customers assigned to week w having a highest frequency equal to f, and let C_d be the total capacity of vehicles during a day d, $d \in D$. We define T_d as a decision variable representing the maximal service time of all clients during each day of planning horizon and x_{id} as a decision variable equal to 1 if client i is visited on day d.

$$\min \quad T_d$$

$$T_d \geq \sum_{i \in C_w} S_{iw} x_{id} \qquad \forall w \in H, \forall d \in D \tag{6}$$

$$C_d \geq \sum_{i \in C_w} Q_{iw} x_{id} \qquad \forall w \in H, \forall d \in D \tag{7}$$

$$\sum_{d \in D} x_{id} = 1 \qquad \forall i \in C \tag{8}$$

$$\sum_{i \in C_{wf}} x_{id} \leq \left\lceil \frac{|C_{wf}|}{|D|} \right\rceil \qquad \forall w \in H, \forall d \in D, \forall f \in \{1, 2, 3, 6, 12\} \tag{9}$$

The first Constraints (6) restrict the total workload of each day. Constraints (7) guarantee that vehicles daily capacity C_d is not exceeded. Constraints (8) make sure that exactly one day is selected for every client. Constraints (9) balance different frequencies of the customers between days of the week.

4.3   Routing Phase

Having assigned all customers to planning days using weeks and days planning models (Sects. 4.1 and 4.2), we need to solve for each day of the planning horizon a variant of VRP, for which specific constraints need to be taken into account. More precisely, in addition to classical VRP constraints such as vehicle capacities, time windows of visits, and limited driving hours per vehicle per day, we need to consider a strong constraint of visiting customers by the same driver. This constraint prevents us from dealing with the VRP for each day of the planning horizon independently.

The objective of routing phase is building optimized routes for each day of the planning horizon in which customers are visited by the same vehicle. This problem is equivalent to selecting a subset of customers to be visited by each vehicle throughout the planning horizon, and for each subset, the route must be optimized in terms of distance, it must not be overloaded on days with high numbers of visits and not underloaded on days with a low number of visits.

In this section, we develop a general framework for solving daily routing problem over planning horizon, while respecting all classical VRP constraints including customer visits by the same vehicle. The framework proceeds as follows:

1. Construct a set of day-types, each containing a list of customers to be visited and requests associated with these customers.
2. Generate routes of each typical day.
3. Project and adjust obtained routes over the days of the planning horizon.

Day-Types Construction. The constraint of having the same customers visited by the same vehicle (agent) prevents us from building routes for each day independently of each other. This first step of solving approach aims to construct day-types so that the routes of each day-type are reused in days of weeks represented by that day-types. More precisely, this step builds a set of d day-types (in our case $d = 5$). Each day-type aggregates all data (customers to be visited, quantities to be served, etc.) of days represented by that day-type, and each day-type corresponds to the same day of the H weeks of the planning horizon. For example, the first day-type contains all data (customers to be visited, etc.) of the first day of every week of the planning horizon. Let $S_{J_k i}$ be the set of customers assigned to the kth day of week i in days planning model (Sect. 4.2), and let $p^j_{J_k i} = \{p^{1,j}_{J_k i}, \ldots, p^{r,j}_{J_k i}\}$ be a set of products requested by customer j, $j \in S_{J_k i}$, and let $q^{1,j}_{J_k i}, \ldots, q^{r,j}_{J_k i}$ be the requested quantities of products $p^{1,j}_{J_k i}, \ldots, p^{r,j}_{J_k i}$, respectively. We define $\tilde{J}_1, \ldots, \tilde{J}_d$ as d day-types, and denote $S_{\tilde{J}_k}$ the set of customers to be visited in the day-type $\tilde{J}_k$, then we set $S_{\tilde{J}_k} = \cup_{i=1}^{H} S_{J_k i}$, and for each customer j of set $S_{\tilde{J}_k}$, we define his set of requested products as $\tilde{p}^j_{\tilde{J}_k i} = \cup_{i=1}^{H} p^j_{J_k i}$ and each product l of set $\tilde{p}^j_{\tilde{J}_k i}$ is requested by a quantity $\tilde{q}_{lj} = \frac{\overline{q}_{lj} + \underline{q}_{lj}}{2}$, where $\overline{q}_{lj} = \max\{q^{l,j}_{J_k i} | i = 1, \ldots, H\}$ and $\underline{q}_{lj} = \min\{q^{l,j}_{J_k i} | i = 1, \ldots, H\}$. At the end of this first step, d day-types are built, and are independent in terms of data (customers to be visited, quantities to be served, etc.) and thus a second phase of routes construction for each day-type can then be launched.

Routes Construction. In this step we develop an ALNS method for constructing routes of each day-type obtained in step Sect. 4.3, with the objective of minimizing the number of used vehicles and the total distance traveled. The classical ALNS algorithm [19] is an iterative process where, at each iteration, a part of the current solution is destroyed and then reconstructed in order to find a better solution. The destruction step consists in removing some nodes from current routes using destroy operators, and placing them into the *unassigned set*. The construction step then inserts the nodes from *unassigned set* into the routes of the partial solution using repair operators. From a set of destroy and repair operators, an adaptive heuristic selects the most successful one to apply in each phase based on a roulette-wheel selection mechanism, favoring the operators that have been successful in recent iterations. This succession of destruction and construction steps are embedded within a local search framework to handle acceptance of new solutions. An outline of proposed ALNS approach is described on Algorithm 1.

Algorithm 1 ALNS outline

    Input: initial feasible solution s
    Output: best feasible solution s^b
1: $s^b \leftarrow s$;
2: $initializeWeights(\rho^-, \rho^+)$;   ▷ ρ^-, ρ^+: Destroy and repair operators weights
3: **repeat**
4:     $selectOperators(\Omega^-, \Omega^+)$;   ▷ Ω^-, Ω^+: Destroy and repair operators vectors
5:     $s^t \leftarrow repair(destroy(s))$;
6:     **if** $accept(s^t, s)$ **then** $s \leftarrow s^t$;
7:     **if** $obj(s^t) < obj(s^b)$ **then** $s^b \leftarrow s^t$;
8:     $updateWeights(\rho^-, \rho^+)$;
9: **until** stop criterion is met;
10: **return** s^b;

Destroy Operators. A destroy operator disconnects q nodes from the current solution, with q being a random number in $[n \cdot \xi_{min}, n \cdot \xi_{max}]$, where n refers to the number of nodes composing the current solution. We used a Random-Removal and a Cluster-Removal operators in our algorithm.

- Random-Removal selects q nodes randomly and disconnects them from the current solution and removes them.
- Cluster Removal, which was introduced in [20], tries to disconnect a large set of related nodes in terms of distance. Cluster Removal starts by choosing one route randomly, then a clustering step is performed. A clustering step consists in partitioning nodes of the current route into two clusters, using a modified Kruskal's algorithm for the minimum spanning tree problem, where algorithm stops when two connected components (clusters) are left in the graph (route). Then one of the two components is selected randomly, and

nodes of this component are disconnected and added to the unassigned set. Another route is selected and the whole process is repeated until q nodes are disconnected.

Repair Operators. A repair operator tries to insert in the current solution the nodes that have been removed and placed in unassigned set together with nodes that could not be inserted in previous solution. This set of nodes is denoted by U. In our method we used the Best-Insertion and the Regret-Insertion operators.

- Best-Insertion. At each iteration, the best insertion cost is computed for each node in the set U and the node with the lowest insertion cost is inserted at its best position. The heuristic stops when U is empty or none of the nodes can be inserted.
- Regret-Insertion. For each node in U $\Delta f_{i,l}$ designates the insertion cost of the node i in the lth best of all routes. At each iteration, the node i^* is selected for the insertion at its best position where $i^* = argmax_{i \in U}(\sum_{l=1}^{k}(\Delta f_{i,l} - \Delta f_{i,1}))$. The heuristic stops when no more unassigned node can be inserted in a route or if U is empty. In this paper, we set k to two.

The Objective function used to evaluate the solution of each iteration of ALNS method has three components: the first concerns total distance traveled by vehicles in the current solution, the second is the number of used vehicles and the third one is the number of unassigned jobs. All these components are weighted by coefficients α, β and γ respectively, and to satisfy problem requirements, these coefficients are set to values where $\alpha \ll \beta \ll \gamma$.

Routes Projection. The constructed routes for day-types allow the same customer to be served by the same vehicle. However, when applying the route-types on weekdays, some customers present in route-types are not planned for visits in some weeks. In such case, unplanned customers are removed from route-types and then the routes are optimized again. The problem of daily intra-route re-optimization corresponds to traveling salesman optimization problem [21], and Lin-Kernighan heuristic [22] is used.

5   Experimental Results

In this section, we present experimental results on the industrial instance. The different algorithms were implemented in Java 8. Mathematical models of first and second stages of solving approach are solved using CPLEX 12.8 in single thread mode, and the ALNS method was implemented in single-threaded sequential mode. Tests are performed on a personal computer with 3.10 GHz Intel Core i7-7920HQ processor and 32 GB RAM running on Windows 10. In our experiment we have not paid attention to the running times of our approach, since the industrial application did not impose limitations on the calculation time used. However, the running times of the different steps of our solving approach are reported.

In the following, detailed characteristics of the industrial instance are provided and experimental results are shown. Note that we cannot solve the problem optimally due to the size of the industrial instance. Furthermore, results of our computational experiment have been analyzed in terms of the objective set at each step of the solving approach and also the final result is compared to the solution already used by the company.

5.1   Instance Description

Here we describe the industrial instance. The company serves 6062 customers, spread out over Paris and Ile-de-France region (see Fig. 1). The planning horizon is 12 weeks and 5 days per week. The company provides a set of 14 types of products, and products are requested with frequencies 1, 2, 3, 6 and 12, where frequency represents the number of visits during the planning horizon. Each customer requests a subset of products and frequency of visits for each product over the planning horizon. In total there are 20813 product-customer requests and when the frequencies are included, the number of requests reaches 69951, which corresponds on average to 11.54 requests per customer. Table 1 gives an overview on the distribution of the number of customers by requested products and Table 2 gives an overview on the distribution of customers by frequency of visits. The average service time of customers is in the interval $[2, 328]$ min. The company uses 40 vehicles of 12 different capacity types (payloads), ranging from 500 Kg to 1600 Kg. There is a central depot where routes start and end. The maximum driving and servicing time is limited to 7 h per day for each driver.

Table 1. Distribution of customers by requested products

# Products	1	2	3	4	5	6	7	8	9	10	11	12	13	14
# Customers	1400	545	1235	1103	1109	328	182	79	43	25	9	1	3	0

Table 2. Distribution of customers by requested frequencies

Frequency	1	2	3	6	12
# Customers	247	369	3403	1259	784

5.2   Weeks Planning Results

The MILP weeks planning model is optimally solved in 8 s. Results are presented in Table 3. Table 3 reports, for each week, the total service time in hours, the number of customers assigned and the maximum distance between two customers in kilometres. The results clearly show a well-balanced total service time between

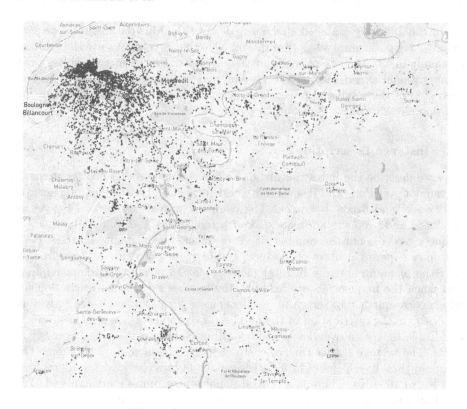

Fig. 1. Overview of customers location

weeks, with a maximum difference of 26 min. However, the number of customers is less balanced between weeks, but this is still a non-essential criterion since customer service times are in a wide range [2, 328] min. It can also be seen that the clients are well distributed geographically since the maximum distances between the clients are almost the same for all weeks.

5.3 Days Planning Results

The MILP days planning model is solved with a gap of 0.2% in 3 min, then no improvement of the solution was observed beyond 30 min. Table 4 reports, for each week and each day of the planning horizon, the total service time in hours, while Table 5 gives, for each day, the distribution of customers over, and the total number of customers assigned to weekdays. The results show a well balanced total service time and the number of customers over days of each week.

Table 3. Results of weeks planning model

Week	1	2	3	4	5	6	7	8	9	10	11	12
Total service time (hours)	547.72	548.13	547.98	548.15	547.83	547.92	547.97	547.98	548.08	548.12	548.08	548.08
#Customers	3120	2308	2137	1987	3043	2240	2171	1839	3051	2386	2123	1823
Maximum distance (km)	86.76	86.90	86.10	86.91	86.76	86.90	86.10	82.59	86.76	87.04	87.76	82.69

Table 4. Service time distribution in days planning model

| | Week | | | | | | | | | | | |
	1	2	3	4	5	6	7	8	9	10	11	12
Monday	109.72	109.50	109.13	109.72	109.23	109.67	109.57	109.50	109.53	109.68	109.60	109.35
Tuesday	109.27	109.77	109.55	109.85	109.42	109.52	109.68	109.87	109.42	109.77	109.58	109.65
Wednesday	109.62	109.82	109.78	109.55	109.57	109.18	109.62	109.82	109.57	109.63	109.67	109.73
Thursday	109.63	109.35	109.68	109.65	109.82	109.78	109.57	109.72	109.87	109.58	109.48	109.75
Friday	109.48	109.70	109.83	109.38	109.80	109.77	109.53	109.08	109.70	109.45	109.75	109.60

Table 5. Customers distribution in days planning model

| | Week | | | | | | | | | | | | Total |
	1	2	3	4	5	6	7	8	9	10	11	12	
Monday	625	462	428	397	610	450	435	369	611	478	426	367	1218
Tuesday	621	462	426	399	608	448	434	365	609	477	424	366	1212
Wednesday	623	461	426	399	608	449	433	368	608	478	425	362	1210
Thursday	626	461	429	398	608	445	434	368	610	476	423	365	1211
Friday	625	462	428	394	609	448	435	369	613	477	425	363	1211

5.4 Routing Results

After constructing day-types from previous assignment results (see Sect. 5.3), we proceed in generating corresponding routes for each day of planning horizon using ALNS algorithm embedded in a simulated annealing (SA) framework as presented in Sect. 4.3. In order to calibrate parameters of our ALNS, we run some preliminary experiments. Some parameters such as adaptive parameters for weight adjustment of destroy and repair operators $(\sigma_1, \sigma_2, \sigma_3)$ are taken from the literature while the remaining parameters are chosen after preliminary experiments as shown in Table 6. Time and distance travel matrices are computed using Google Distance Matrix API [23].

First, we solved the five day-types instances resulting from the assignment of customers to day-types (Sect. 5.3). For each day-type instance, we restricted the route duration limit to 7 h and 15 min per driver as these instances include all customers of specific days of planning horizon. A computing time limit of 2 h was

Table 6. Parameter setting of ALNS

Parameters	$[\xi_{min}, \xi_{max}]$	SA init. temper.	SA cooling rate temper.	$(\sigma_1, \sigma_2, \sigma_3)$	(α, β, γ)
Values	$[0.01, 0.05]$	200	0.9996	$(33, 9, 13)$	$(1, 3000, 120000)$

set for all instances. Ten test runs per instance were performed, showing a very small difference in final results. Best results are reported in Table 7 showing, for each day, the number of used vehicles, the total distance traveled in kilometres and the average route duration per vehicle. We observe that the number of vehicles used in each day-type is the same and the number of kilometres is well balanced between day-types.

Table 7. Day-types routing results

Days	#Customers	#Used vehicles	Total distance (km)	Average route duration (hours)
Monday	1218	35	2606.38	7.16
Tuesday	1212	35	2508.65	7.19
Wednesday	1210	35	2697.38	7.20
Thursday	1211	35	2569.46	7.15
Friday	1211	35	2433.69	7.16

The last step in the routing phase was re-optimizing route-types for each day of the planning horizon as described in Sect. 4.3. Tables 8 and 9 report total distances and the average routing duration obtained for each day of every week, respectively. For each weekday, the average distance over all weeks varies between 75.1% and 90.4% of the total distances calculated for day-types whereas the average routing time varies between 63,1% and 74,5% of the average time calculated for day-types. These gaps are explained by the absence of some clients, considered in day-types, in the actual sets of clients to be served in each day of planning horizon. To improve those results, we tried to increase values of route duration limit set actually to 7 h and 15 min in day-types routing generation, and we observe that a higher route duration limit leads to some customer demands which cannot be inserted in real instances as they violate the 7-h limit of drivers.

The company uses 40 vehicles in their actual solution, so we obtain a gain of 5 vehicles. Note that our results are strongly dependent with choosing day-types as an approach to tackle the problem, especially with the constraint stating that customers must always be served by the same agent. We believe that current solution can still be improved in terms of number of used vehicles, as the average routing duration has not reached yet the 7-h limit.

Table 8. Total distances obtained for real instances over weeks in kilometres

	Weeks											
	1	2	3	4	5	6	7	8	9	10	11	12
Monday	2257.6	2226.8	2148.5	2152.5	2251.2	2188.1	2191.3	2036.2	2241.7	2267.8	2176.7	2046.2
Tuesday	2052.6	2053.9	2036.9	2061.4	2051.3	2058.2	2032.1	1997.7	2048.4	2162.6	2037.9	1998.5
Wednesday	2439.3	2263.8	2185.3	2146.1	2414.2	2321.9	2219.0	2025.8	2414.9	2299.0	2160.4	2120.5
Thursday	2297.5	2182.7	2005.0	2080.8	2270.2	2183.1	2012.5	1971.4	2271.4	2223.3	2000.1	2002.2
Friday	2188.9	1994.8	2016.0	1892.7	2179.9	1983.2	2022.9	1934.2	2152.8	2041.3	2018.5	1858.9

Table 9. Average routing durations obtained for real instances over weeks in hours

	Weeks											
	1	2	3	4	5	6	7	8	9	10	11	12
Monday	5.27	5.09	5.04	4.73	5.27	4.97	5.14	4.73	5.22	5.05	5.04	4.79
Tuesday	5.39	5.22	5.09	4.90	5.33	5.16	5.10	4.82	5.29	5.21	5.14	4.92
Wednesday	5.37	5.12	4.78	4.85	5.34	5.18	4.79	4.70	5.35	5.09	4.78	4.55
Thursday	5.26	5.06	4.98	4.94	5.15	4.93	4.93	4.69	5.30	5.11	4.79	4.72
Friday	5.27	5.03	4.92	4.93	5.32	5.08	4.97	4.83	5.30	5.12	5.00	4.81

6   Conclusion

In this paper, we have investigated the design of tactical plans for a routing problem involved in a company of hygiene services. We modeled the problem as multiple periods VRP and proposed a decomposition approach, using integer programming models and ALNS method. We were able to improve the current routes of company using 5 vehicles less over the planning horizon. In order to improve the solving approach, future research will include more sophisticated optimization techniques for weeks and days planning models that consider distances between customers, and improve the routing phase to reach the restricted limit duration of routes.

References

1. Nuortio, T., Kytöjoki, J., Niska, H., Bräysy, O.: Improved route planning and scheduling of waste collection and transport. Expert Syst. Appl. **30**(2), 223–232 (2006)
2. Matos, A.C., Oliveira, R.C.: An experimental study of the ant colony system for the period vehicle routing problem. In: Dorigo, M., Birattari, M., Blum, C., Gambardella, L.M., Mondada, F., Stützle, T. (eds.) ANTS 2004. LNCS, vol. 3172, pp. 286–293. Springer, Heidelberg (2004). https://doi.org/10.1007/978-3-540-28646-2_26
3. Coene, S., Arnout, A., Spieksma, F.C.R.: On a periodic vehicle routing problem. J. Oper. Res. Soc. **61**(12), 1719–1728 (2010)

4. Ronen, D., Goodhart, C.A.: Tactical store delivery planning. J. Oper. Res. Soc. **59**(8), 1047–1054 (2008)
5. An, Y.J., Kim, Y.D., Jeong, B.J., Kim, S.D.: Scheduling healthcare services in a home healthcare system. J. Oper. Res. Soc. **63**(11), 1589–1599 (2012)
6. Blakeley, F., Argüello, B., Cao, B., Hall, W., Knolmajer, J.: Optimizing periodic maintenance operations for Schindler elevator corporation. Interfaces **33**(1), 67–79 (2003)
7. Campbell, A.M., Wilson, J.H.: Forty years of periodic vehicle routing. Networks **63**(3), 276–276 (2014)
8. Beltrami, E.J., Bodin, L.D.: Networks and vehicle routing for municipal waste collection. Networks **4**(1), 65–94 (1974)
9. Russell, R., Igo, W.: An assignment routing problem. Networks **9**(1), 1–17 (1979)
10. Christofides, N., Beasley, J.E.: The period routing problem. Networks **14**(2), 237–256 (1984)
11. Russell, R.A., Gribbin, D.: A multiphase approach to the period routing problem. Networks **21**(7), 747–765 (1991)
12. Cordeau, J.F., Gendreau, M., Laporte, G.: A tabu search heuristic for periodic and multi-depot vehicle routing problems. Netw. Int. J. **30**(2), 105–119 (1997)
13. Cordeau, J.F., Laporte, G., Mercier, A.: A unified tabu search heuristic for vehicle routing problems with time windows. J. Oper. Res. Soc. **52**(8), 928–936 (2001)
14. Francis, P., Smilowitz, K., Tzur, M.: The period vehicle routing problem with service choice. Transp. Sci. **40**(4), 439–454 (2006)
15. Alegre, J., Laguna, M., Pacheco, J.: Optimizing the periodic pick-up of raw materials for a manufacturer of auto parts. Eur. J. Oper. Res. **179**(3), 736–746 (2007)
16. Mourgaya, M., Vanderbeck, F.: Column generation based heuristic for tactical planning in multi-period vehicle routing. Eur. J. Oper. Res. **183**(3), 1028–1041 (2007)
17. Hemmelmayr, V.C., Doerner, K.F., Hartl, R.F.: A variable neighborhood search heuristic for periodic routing problems. Eur. J. Oper. Res. **195**(3), 791–802 (2009)
18. Vidal, T., Crainic, T.G., Gendreau, M., Lahrichi, N., Rei, W.: A hybrid genetic algorithm for multidepot and periodic vehicle routing problems. Oper. Res. **60**(3), 611–624 (2012)
19. Ropke, S., Pisinger, D.: An adaptive large neighborhood search heuristic for the pickup and delivery problem with time windows. Transp. Sci. **40**(4), 455–472 (2006)
20. Ropke, S., Pisinger, D.: A unified heuristic for a large class of vehicle routing problems with backhauls. Eur. J. Oper. Res. **171**(3), 750–775 (2006)
21. Bellmore, M., Nemhauser, G.L.: The traveling salesman problem: a survey. Oper. Res. **16**(3), 538–558 (1968)
22. Lin, S., Kernighan, B.W.: An effective heuristic algorithm for the traveling-salesman problem. Oper. Res. **21**(2), 498–516 (1973)
23. Google: Distance matrix API. https://developers.google.com/maps/documentation/distance-matrix/start. Accessed 30 Oct 2018

Rigorous Performance Analysis
of State-of-the-Art TSP Heuristic Solvers

Paul McMenemy[1](✉)(iD), Nadarajen Veerapen[2](iD), Jason Adair[1](iD),
and Gabriela Ochoa[1](iD)

[1] Computing Science and Mathematics, University of Stirling, Stirling, UK
paul.mcmenemy@stir.ac.uk
[2] Univ. Lille, CNRS, Centrale Lille, UMR 9189 - CRIStAL, 59000 Lille, France

Abstract. Understanding why some problems are better solved by one algorithm rather than another is still an open problem, and the symmetric Travelling Salesperson Problem (TSP) is no exception. We apply three state-of-the-art heuristic solvers to a large set of TSP instances of varying structure and size, identifying which heuristics solve specific instances to optimality faster than others. The first two solvers considered are variants of the multi-trial Helsgaun's Lin-Kernighan Heuristic (a form of iterated local search), with each utilising a different form of Partition Crossover; the third solver is a genetic algorithm (GA) using Edge Assembly Crossover. Our results show that the GA with Edge Assembly Crossover is the best solver, shown to significantly outperform the other algorithms in 73% of the instances analysed. A comprehensive set of features for all instances is also extracted, and decision trees are used to identify main features which could best inform algorithm selection. The most prominent features identified a high proportion of instances where the GA with Edge Assembly Crossover performed significantly better when solving to optimality.

Keywords: TSP · Algorithm selection · EAX · GPX ·
Performance analysis

1 Introduction

Despite decades of intense study, the Travelling Salesperson Problem (TSP) sustains its practical and theoretical interest. It has inspired the design of powerful exact and heuristic solvers, able to tackle TSP problems of increasing size in shorter computing time. The objective in the TSP, given a set of n locations (generally called cities) and pairwise distances between them, is to find the shortest round-trip through all cities such that the total length of the trip (a tour) is minimised. Here we consider the most common case of the problem, the 2D symmetric TSP, where cities correspond to points in the Euclidean plane and distances are also Euclidean. The current TSP state-of-the-art exact solver, Concorde [1], remains unbeaten. Concorde has been used to optimally solve instances

© Springer Nature Switzerland AG 2019
A. Liefooghe and L. Paquete (Eds.): EvoCOP 2019, LNCS 11452, pp. 99–114, 2019.
https://doi.org/10.1007/978-3-030-16711-0_7

of several thousand cities and, for fewer than 1 000 cities, does so in very feasible running times. However, there is interest in developing inexact or heuristic solvers, as they can provide surprisingly good results for large instances in reasonably short amounts of time when compared to obtaining a solution via Concorde. The scenario of heuristic solvers was previously dominated by a single contender for several years: Helsgaun's Lin-Kernighan Heuristic (LKH+IPT) [2,3]. However, recent evolutionary algorithms using the Edge Assembly Crossover (EAX) [4], as well as hybrid approaches using the Generalised Partition Crossover in concert with LKH (LKH+GPX2), have been shown to match and improve upon LKH+IPT performance in some instances [5].

An outstanding challenge in heuristic optimisation is to understand how to find the most suitable algorithm for a given problem instance or set of instances. Corne and Reynolds [6] introduce the notion of the 'footprint' of an algorithm to indicate how its performance generalises across different dimensions of instance space. Smith-Miles and Lopes [7] followed this by proposing a methodology to determine the relative performance of optimisation algorithms across various classes of instances. Later works by Pihera and Musliu [8], Kotthoff [9], and Kerschke et al. [10] show that per-instance automated algorithm selection techniques can be used to improve the state-of-the-art in inexact TSP solving.

The main goal of this study is to perform rigorous tests comparing the runtimes to optimality of the three previously mentioned TSP heuristic solvers: (i) LKH+IPT; (ii) an evolutionary algorithm with EAX; and (iii) a hybrid evolutionary algorithm combining LKH with GPX2. A comprehensive set of features is then to be used to characterise TSP instances taken from the available benchmark sets and instance generators in the literature, with the aim of identifying specific instance space features which can provide guidance on which solver is most effective in solving to optimality.

2 Methodology

2.1 Instances

For this study 1800 symmetric TSP instances were generated, comprised of varying instance sizes and structures:

Random Uniform Euclidean (RUE): instances generated by placing a number of points (representing the cities to be visited) randomly within a planar square (e.g., Fig. 1(a)). The distances between the cities are determined as the Euclidean distances between the respective points, where the cost of travel between cities is specified as the Euclidean distance rounded to the nearest whole number. Instances were generated using the DIMACS `Portgen` generator[1] with sizes $n \in \{500, 1\,000, 1\,500, 2\,000\}$. One hundred and fifty seeds of each instance size n were generated for this study.

---

[1] http://archive.dimacs.rutgers.edu/Challenges/TSP/.

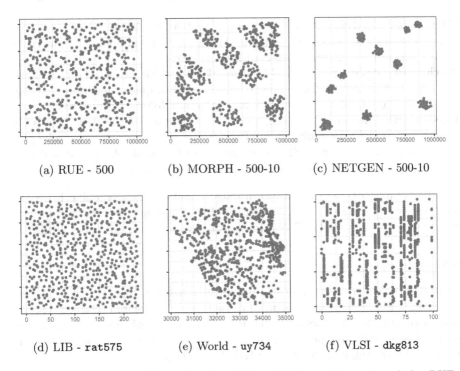

(a) RUE - 500   (b) MORPH - 500-10   (c) NETGEN - 500-10

(d) LIB - rat575   (e) World - uy734   (f) VLSI - dkg813

Fig. 1. Instance type problem domains. Plots (a)–(c) are examples of the RUE, MORPH & NETGEN instances, all with $n = 500$. The MORPH example, (b), is constructed from (a) and (c), with $\alpha = 0.5$. Plot (d) shows rat575, a rattled grid of 575 locations; plot (e) is the World instance with 734 town/cities in Uruguay (uy734); and (f) is the VLSI problem denoted as dkg813

Random Clustered Euclidean (NETGEN): a set of instances generated with sizes $n \in \{500, 1\,000, 1\,500, 2\,000\}$. Each instance was created with a corresponding parameter, $c \in \{2, 5, 10\}$, which specifies the number of clusters located within the instance by latin hypercube method. City locations are then distributed with respect to the cluster centres, maintaining cluster segregation, Fig. 1(c). Instances were created using the netgen software package within the open source software R [11, 12]. Again, 150 instances were generated for each size n, split into 50 instances per combination of c and n.

Morphed Euclidean (MORPH): instances which have been generated from a combination of equal n-sized RUE and NETGEN instances. Pairs of RUE (x_i) and NETGEN (y_i) cities are greedily matched, the first pair (x_1, y_1) being the closest by euclidean distance, with this repeated $\forall i \in \{2, 3, ..., n\}$ until all cities have been matched. Each MORPH city is then generated by relocating a proportional distance (defined by a parameter α) along a straight line from x_i to y_i. Figure 1(b) shows a MORPH instance, generated from a 500-city RUE instance, Fig. 1(a), and 500 city NETGEN instance with 10 clusters, Fig. 1(c). The MORPH examples in this study were generated using the TSPMETA package [13] and with $\alpha = 0.5$.

LIB: a widely-used collection of instances with different characteristics and TSP applications [14]. Instances with size $400 \leqslant n < 5\,000$ and edge weight types EUC 2D & CEIL 2D were selected for this study.

World: instances from the World benchmark set[2] which are based on real locations of cities in different countries. Cities with duplicate coordinates were not considered, all instances with size $400 < n < 5\,000$ were selected.

VLSI: a set of instances that originate from an application in very large scale integration (VLSI) circuit design. These instance types are known to be particularly hard for many TSP solvers, including Concorde. Again, all instances with sizes $400 < n < 5\,000$ were included in this study [15].

Only 93 instances of the required size and edge weight types are available, hence the requirement to generate a large, supplemental set of instances. Where appropriate, the combined instances of LIB, World and VLSI will be referred to as TSPLIB.

2.2 Solvers

Our study considers three high-performance heuristic solvers for the symmetric TSP. The three solvers incorporate a form of recombination (crossover) and, according to the literature [10, 16], achieve state-of-the-art performance on Euclidean TSP instances. The first two solvers are modern variants of the powerful Lin-Kernighan-Helsgaun (LKH) heuristic [3].

LKH+IPT & LKH+GPX2: Multi-try LKH variants. LKH is an iterated local search algorithm based on the the Lin-Kernighan heuristic (LK) [17], a variable-depth method that generates complex local search moves by heuristically constructing a sequence of edge exchanges. Over the years, several improvements have been incorporated to LKH. The best results in the literature have been obtained with a version known as multi-trial LKH, where solutions generated by soft restarts of the LK heuristic are recombined using Iterative Partial Transcription (IPT). An alternative version of multi-trial LKH has been recently proposed [16], where the IPT recombination operator is substituted by GPX2, a new generalised partition crossover proposed for the TSP. Both IPT and GPX2 are forms of partition crossover, which are deterministic recombination operators that use features common to the parents to decompose the evaluation function. Following [16] we consider these two versions of multi-try LKH, and name them as LKH+IPT and LKH+GPX2.

EAX: GA with Edge Assembly Crossover. A series of high-performing evolutionary algorithms for the TSP integrate variants of edge assembly crossover, a recombination operator that combines the edges of two parent solutions trying to add only few, short edges not found in any of the two parents [18]. We consider the most recent version of EAX [4], which aggregates three key enhancements: (i) initialisation of the population by local optimisation, (ii) improved local and global variants of the edge assembly crossover operator, and (iii) specific diversity preservation techniques that use edge entropy measures in the population replacement scheme.

---

[2] http://www.math.uwaterloo.ca/tsp/world/countries.html.

2.3   Experimental Setup

Each of the solvers are implemented as single-threaded programs and were run on Intel Xeon Gold 6138 2.0 GHz CPUs[3]. The common termination criterion applied is one hour of CPU time. For the analyses in this study, the PAR10 penalised runtime was implemented [10]. This assigns a penalized runtime of 10 times the termination criterion to runs that fail to solve to optimality within that limit. Each solver is run 30 times on each instance and times are recorded.

For EAX, the parameters used are the ones prescribed by Nagata and Kobayashi [4] for "small" instances ($n < 10\,000$) in the `readme` file accompanying their source code. The population size is therefore set to 100 and the number of offspring set to 30. For LKH, version 2.0.9 is used and the default parameters specified within the source code and the example configuration file are considered. The `RECOMBINATION` parameter is set to either `IPT` or `GPX2` in order to select the appropriate crossover operator.

Neither EAX nor LKH implement a time-based termination criterion but use a fixed number of trials instead. The source code was therefore modified to run within a specific amount of time and allow for consistent comparison between EAX and the two LKH variants.

In order to know whether a solver was successful, the result of each run is compared to the known optimal objective function values for library instances. For generated instances, the state-of-the-art exact TSP solver Concorde [1,19] was used to generate an optimal solution whose objective function value is used for comparison.

2.4   Instance Features

Two sets of features were calculated and combined together for use in this study:

TSPMETA: a set of features described by Mersmann et al. [13] which provide a group of standard geometric features derived from TSP instances. The features were calculated using the `TSPMETA` package within the R software package [12], with 64 features generated in total.

Pihera: a feature set defined by Pihera and Musliu [8] which are based on kNN graph transformations of each instance and the generation of extensive summary statistics of the kNN graphs. This generated 285 features, mainly comprised of kNN-graph transformation metrics and their summary statistics.

Pihera features were calculated for each of the instance type sets described earlier, with no group taking longer than 30 s using the openly available C++ software[4]. The `TSPMETA` features took noticeably longer to calculate, roughly doubling in execution time as sizes incremented by 500 cities. Both the TSPMETA and the Pihera features were combined to generate a comprehensive feature set of 349 features. This combined group of features improved the likelihood of extracting feature(s) which could best inform heuristic selection.

---

[3] https://www.archie-west.ac.uk.

[4] https://tspalgsel.github.io/#software.

2.5   Statistical Evaluation of Heuristic Performance

Performance analysis was carried out on the three state-of-the-art solvers described earlier (LKH+IPT, EAX, LKH+GPX2). A round-robin set of statistical comparisons was carried out on pairs of heuristics: i.e., EAX versus LKH+IPT, LKH+IPT versus LKH+GPX2, and LKH+GPX2 versus EAX. Mann-Whitney statistical tests were carried out on these pairs for each TSP instance: the thirty runs of each heuristic pair were tested against each other to determine whether their respective runtimes could be from the same distribution of values. This was carried out at the 99% confidence level, thus ensuring a high level of likelihood that runtimes were identified as being from differing distributions if the Mann-Whitney tests' p-values < 0.01. When this occurred, a further comparison of each heuristic's median runtimes was carried out, and the solver with the lower median runtime value was deemed to be the best of the two tested.

The null hypothesis of the Mann-Whitney test posits that the runtimes of the two solvers being tested against one another are from the same distribution of values and so, if p-values > 0.01, then the null hypothesis could not be rejected. In this circumstance it was deemed that neither heuristic was significantly better than the other at solving the instance to optimality, and so neither was designated as the best heuristic for the particular instance being tested. This outcome was labelled as "ANY" in the following results.

3   Performance Analysis

Runtime performance of each heuristic was carried out using the methodology described in the previous section. Here, results are presented by pairwise combination, i.e., one heuristic is measured against another heuristic's performance. Firstly, scatter plots of the median runtimes per instance are presented, with plots further subdivided by the instance types NETGEN, MORPH, RUE and TSPLIB included in this study. In all plots, red scatter points correspond to when EAX performed significantly better, blue points to LKH+GPX2, and green to LKH+IPT. Black scatter points denote instances where one heuristic was not deemed to be significantly quicker than the other. Diagonal black lines on each plot indicate equality of median runtimes; boxplots of median runtime values are also provided, located adjacent to opposite axes to provide additional comparison of performance.

For each paired comparison, tables showing the numbers and percentages of times of classifications are also presented, with the TSPLIB instances broken down by VLSI, World and LIB instance types. The tables provide further detail that can be occluded in scatter plots by coincidental or clustered data points.

3.1   EAX v LKH+IPT Performance Results

Firstly, the relative performances of EAX versus LKH+IPT are considered, and results are shown in Fig. 2 and Table 1. It can be seen from Fig. 2(a) and (b) that EAX is vastly superior in solving NETGEN and MORPH instance types to optimality when compared with LKH+IPT. The median value of the boxplot for EAX in Fig. 2(a) (which is the overall median value of all EAX runs for NETGEN) is 7.2 s, whereas the corresponding value for LKH+IPT is 107.7 s. Similar values are returned for the MORPH instances, with median EAX runtime = 7.6 s and median LKH+IPT = 116.4 s.

However, for the RUE instances, Fig. 2(c) shows that LKH+IPT greatly improves performance when compared against EAX. Results shown in Table 1 reinforce this result, showing that LKH+IPT solved 41.5% of instances significantly quicker than EAX (median LKH+IPT runtime = 9.0 s). Conversely, EAX only solved 23.8% of RUE instances better than LKH+IPT (median EAX runtime 10.6 s).

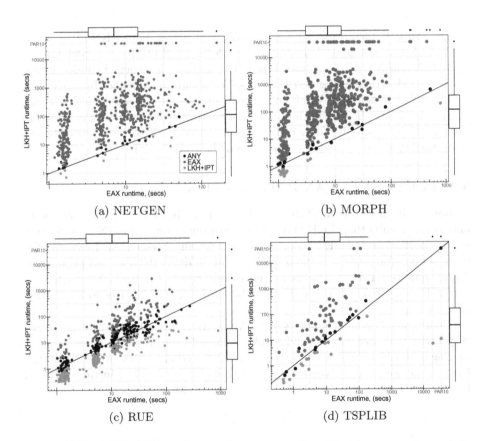

(a) NETGEN

(b) MORPH

(c) RUE

(d) TSPLIB

Fig. 2. EAX v LKH+IPT runtime performance results. All plot axes are log-scaled; scatter points are median values of 30 runs of pairwise instances in the form (median(EAX), median(LKH+IPT)); diagonal lines represent equality of median performance; boxplots on axes show distribution of median runtime values (Color figure online)

Table 1. EAX versus LKH+IPT classifications

Instance type	EAX	LKH+IPT	ANY
RUE	143 (23.8%)	249 (41.5%)	208 (34.7%)
MORPH	562 (93.7%)	16 (2.7%)	22 (3.7%)
NETGEN	567 (94.5%)	11 (1.8%)	22 (3.7%)
TSPLIB:			
VLSI	38 (66.7%)	8 (14.0%)	11 (19.3%)
World	3 (100.0%)	0 (0.0%)	0 (0.0%)
LIB	18 (54.5%)	8 (24.2%)	7 (21.2%)
Total	1331 (70.3%)	292 (15.4%)	270 (14.3%)

For the LIB instances, EAX still performed better than LKH+IPT; however, EAX did not overwhelmingly outperform LKH+IPT as it had for the NETGEN and MORPH instance types. The 3 World instances included in our study (uy734, zi929 and mu1979) were solved significantly better by EAX than LKH+IPT. Focusing on VLSI only, the median runtime value for EAX was 12.8 s, with LKH+IPT returning a median runtime of 59.2 s. When considering these with the results in Table 1, it is apparent that EAX solves VLSI instances to optimality with noticeably greater speed than LKH+IPT.

3.2   LKH+GPX2 v EAX Performance Results

Examining both Fig. 3 and Table 2, it is apparent that EAX provides considerably better results than LKH+GPX2. As with the results for EAX v LKH+IPT in the previous section, EAX performs substantially better than LKH+GPX2, most notably for NETGEN (95.0% EAX) and MORPH (93.5% EAX) instances. Again, however, EAX performs less effectively versus LKH+GPX2 when solving RUE types; EAX drops to 25.8%, whereas LKH+GPX2 rises to 38.8% best effectiveness.

Comparing the classifications for EAX across Tables 1 and 2, there exists very little difference between all results, thus inferring that EAX performs as well against LKH+IPT as it does versus LKH+GPX2. Also, note again that EAX is the dominant solver for VLSI, with 64.9% of all VLSI instances solved best by EAX, and only 7 out of 57 best solved by LKH+GPX2.

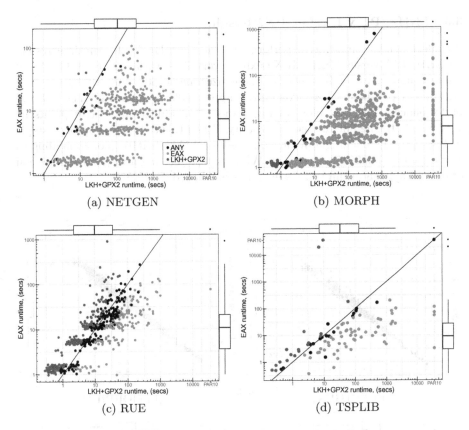

(a) NETGEN   (b) MORPH

(c) RUE   (d) TSPLIB

Fig. 3. LKH+GPX2 v EAX runtime performance results. All plot axes are log-scaled; scatter points are median values of 30 runs of pairwise instances (LKH+GPX2, EAX); diagonal lines represent equality of median performance; boxplots on axes show distribution of median runtime values (Color figure online)

Table 2. LKH+GPX2 versus EAX classifications

Instance type	LKH+GPX2	EAX	ANY
RUE	233 (38.8%)	155 (25.8%)	212 (35.3%)
MORPH	14 (2.3%)	561 (93.5%)	25 (4.2%)
NETGEN	10 (1.7%)	570 (95.0%)	20 (3.3%)
TSPLIB:			
VLSI	7 (12.3%)	37 (64.9%)	13 (22.8%)
World	3 (100.0%)	0 (0.0%)	0 (0.0%)
LIB	9 (27.3%)	19 (57.6%)	5 (15.2%)
Total	276 (14.6%)	1342 (70.9%)	275 (14.5%)

3.3 LKH+IPT v LKH+GPX2 Results

As inferred by their relative performances against EAX, the runtime results
for LKH+IPT versus LKH+GPX2 exhibit very little variation. Both Fig. 4 and
Table 3 show that, for instances of these sizes and types, and with a stopping
criterion of one hour, there exists little appreciable difference in their relative
performances. The four plots in Fig. 4 show that almost all scatter points lie
very closely to the line of median equality. When this did not hold, for example
in Fig. 4(d) where a single point is located at (1329, PAR10) (corresponding to
VLSI instance bck2217), the Mann-Whitney test does not always allow rejection
of its null hypothesis and so the instance is labelled as ANY.

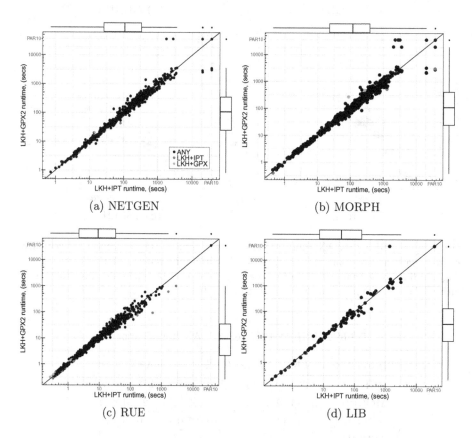

Fig. 4. LKH+IPT v LKH+GPX2 runtime performance scatter plots

Table 3 further emphasises the similarities of runtimes for both LKH+IPT
and LKH+GPX2. 97.8% of the 1893 instances studied were classed as being best
solved by either heuristic: only the RUE types returned any notable numbers of
instances solved better by either solver.

Table 3. LKH+IPT versus LKH+GPX2 classifications

Instance type	LKH+IPT	LKH+GPX2	ANY
RUE	12 (2.0%)	17 (2.8%)	571 (95.2%)
MORPH	4 (0.6%)	3 (0.5%)	593 (98.8%)
NETGEN	1 (0.1%)	3 (0.5%)	596 (99.3%)
TSPLIB:			
VLSI	0 (0.0%)	0 (0.0%)	57 (100.0%)
World	0 (0.0%)	0 (0.0%)	3 (100.0%)
LIB	0 (0.0%)	2 (6.5%)	31 (93.9%)
Total	17 (0.9%)	25 (1.3%)	1851 (97.8%)

3.4 Overall Performance Classification

Due to the overall dominance of EAX in these results, and the similar behaviours of LKH+IPT and LKH+GPX2 (now referred to jointly as SET-LKH), an appropriate question to answer to determine the best solver is "When should we not apply EAX?". If this can be answered, then three courses of action are available: (i) Use EAX only; (ii) use one of SET-LKH; (iii) use any of the above. Due to the close similarity between the SET-LKH results, only the results from the EAX v LKH+IPT and EAX v LKH+GPX2 comparisons were used to determine the overall best heuristic per instance. For overall classification, it is assumed that LKH+IPT and LKH+GPX2 did not perform significantly differently for any instances; only 2.2% of instances were solved significantly more quickly by one than the other (cf. Table 3), and thus is a reasonable simplification to make which does not adversely affect best solver selection.

Table 4 summarises the classification permutations (excluding LKH+IPT v LKH+GPX2) and their outcomes which are used to determine the best overall heuristic per instance. Each permutation of EAX v LKH+IPT and EAX v LKH+GPX2 results in different hierarchical results which are shown as pictograms in Table 4; LKH+IPT denoted by a green "I", EAX (red "E") and LKH+GPX2 (blue "G").

EAX was deemed the best heuristic if it was significantly better than both SET-LKH solvers, or better than one while not defeated by the other (top 3 rows of Table 4). The same reasoning applies for SET-LKH (middle 3 rows), whereas the ANY class was applied only if EAX was not significantly better than both SET-LKH solvers, and vice versa. The permutations where EAX defeated one of the SET-LKH solvers but was defeated by the other did not occur in our study, and so is not shown here. Applying this methodology we can determine whether EAX, one of the SET-LKH heuristics, or ANY of the three would be the best approach.

Table 5 shows that, out of the 1893 instances tested, 1361 were solved significantly better by EAX. EAX also comprises part of the ANY category, and so can be identified as being the most effective heuristic, or at least one of the most effective, for 84.1% of the instances tested.

Table 4. Best heuristic classification by instance

EAX v LKH+IPT	EAX v LKH+GPX2	Pictogram	Best Heuristic
EAX	EAX	E_{GI}	EAX
EAX	ANY	EG_I	EAX
ANY	EAX	$E_G{}^I$	EAX
LKH+IPT	ANY	EG^I	SET-LKH
ANY	LKH+GPX2	$E^G{}_I$	SET-LKH
LKH+IPT	LKH+GPX2	E^{GI}	SET-LKH
ANY	ANY	EGI	ANY

Excluding the RUE instances, the proportion best solved by the classifications EAX or ANY rises to 96.6%. Conversely, considering only the RUE instances, we see that the SET-LKH heuristics perform effectively 72.5% of the time when combined with the ANY classification. The VLSI instances, often considered the most intractable or difficult to solve to optimality, are best solved by EAX 68.4%; only 14.0% are solved significantly quicker by SET-LKH solvers.

Table 5. Best overall heuristic classifications

Type	EAX	SET-LKH	ANY
RUE	165 (27.5%)	257 (42.8%)	178 (29.7%)
MORPH	563 (93.8%)	16 (2.7%)	21 (3.5%)
NETGEN	572 (95.3%)	11 (1.8%)	17 (2.8%)
TSPLIB:			
VLSI	39 (68.4%)	8 (14.0%)	10 (17.5%)
World	3 (100.0%)	0 (0.0%)	0 (0.0%)
LIB	19 (57.5%)	9 (27.3%)	5 (15.2%)
Total	1361 (72.9%)	301 (15.9%)	231 (12.2%)

The marked variations in these results show how important it is to understand the underlying TSP instance itself, and the features of the instance. Doing so can provide important guidance on which solver can be implemented most effectively.

4 Heuristic Selection by Minimal Feature Extraction

It is important to be able to identify instance features which can inform specific algorithm selection. To this end, a decision tree model was trained using the Pihera and TSPMETA feature sets described earlier. A decision tree is a simple but useful technique for creating "rules of thumb" that can be applied to identify instance traits that may hamper the performance of specific solvers. This process identified key features (and their associated values) which can be used to select the most effective instance-specific algorithms. From the head of the decision tree shown in Fig. 5, two key features were extracted which can be used to identify a significant variance in performance between the EAX and SET-LKH solvers: nn5.sc.max.n, from the Pihera feature set; and mst_dists_median, from the TSPMETA feature set.

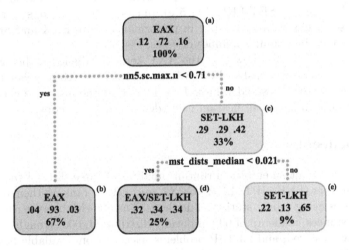

Fig. 5. Decision tree trained on all instance features (tree depth = 2)

It can be seen that 72% of instances were solved significantly faster by EAX than either of the SET-LKH heuristics and, in a further 16%, no significant difference was found (cf. Table 5 and Fig. 5(a)). The intuitive choice, therefore, would be to use EAX by default; however, note that 12% of all instances tested were best solved by the SET-LKH heuristics. It was found that 67% of all instances exhibited nn5.sc.max.n < 0.71; of these, 93% were solved best by EAX (cf. Figs. 5(b) and 6(a)). When nn5.sc.max.n >= 0.71, 71% were solved by SET-LKH heuristics in comparable (29%) or less (42%) time, Fig. 5(c).

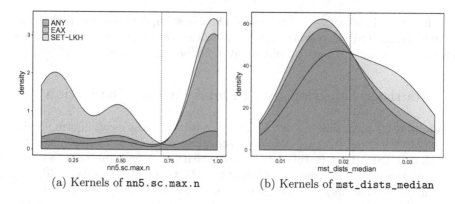

(a) Kernels of nn5.sc.max.n   (b) Kernels of mst_dists_median

Fig. 6. Plots show kernels of ANY, EAX, SET-LKH classed instances for features nn5.sc.max.n (a) and mst_dists_median (b). Vertical lines indicate the threshold values obtained from the decision tree in Fig. 5 (Color figure online)

This branch of the decision tree can be further broken down when feature mst_dists_median $>= 0.021$, with 87% of instances on that branch solved in similar or less time by SET-LKH, Fig. 5(e). In cases where mst_dists_median < 0.021, we see a less distinctive split in performance, with EAX and SET-LKH performing best in a similar number of cases, Fig. 5(d).

Figures 6(a) and (b) show that the ANY kernels, labelled as being blue in colour, are completely masked by the EAX and SET-LKH kernels either side of the decision tree threshold values. This infers that the instances classified as ANY may defy simple identification using decision tree methods.

5   Conclusions

We have carried out an empirical runtime analysis of three state-of-the-art TSP heuristic solvers (LKH+IPT, LKH+GPX2, and EAX), ranking their respective performances using robust statistical methods. The algorithms were run on 1893 distinct instances, made up of 600 uniformly distributed, 600 strongly clustered, 600 loosely clustered, and 93 TSP problems selected from available benchmark datasets. Our results show that EAX performs significantly better than the others, especially for the NETGEN and MORPH instances with almost all of those types solved significantly better by EAX. However, for the RUE types, EAX does not perform as well as the SET-LKH solvers: EAX was only classed as the best solver ∼25% of the time.

EAX was ranked the best overall solver per instance, identified as being significantly better in 72.9% of instances. Including the ANY category (instances where no heuristic proved to be significantly better), EAX could be applied to 85% of all instances tested to return the quickest time to optimality. Even though we observed EAX as being strongly dominant, these results further support the "No Free Lunch" theorem [20], i.e., that no heuristic outperforms any other solver for all problems.

Best algorithm selection is a topic of ongoing research across all combinatorial optimisation problems [8,9,21]. For the TSP instance set analysed here, we have shown that the simple use of a decision tree can allow identification of a few discriminatory features of the instance space that pinpoint which algorithm is best applied. Our analysis identified one feature from the Pihera set (nn5.sc.max.n) whose value strongly predicts when to use EAX. When this is used in concert with another feature from the TSPMETA group (mst_dists_median), we can quickly identify when to apply one of the SET-LKH solvers. Thus, calculating both TSPMETA and Pihera features sets of the TSP instances can provide strong guidance to inform which solver to use.

There is no evidence to assume that EAX would continue to remain dominant over the SET-LKH solvers for increased instance sizes, or that the similar results for both LKH+IPT and LKH+GPX2 would also hold. Therefore an extension of the work presented here would be to carry out similar analyses for symmetric TSP instances of size >2 000. It is anticipated that, due to the effects of combinatorial explosion, a similar study of increased instance size would need to implement a timeout criterion greater than the one hour used here. Renato et al. [16] indicate that LKH+GPX2 does outperform LKH+IPT for most instances they studied which had sizes in the range [3056, 115475]; however, their study was not limited by a timeout criterion, and instead enforced a minimum number of trials per run. Thus, increasing the timeout criterion for larger instances may allow the LKH+GPX2 heuristic the capacity to realise any advantage it may have over LKH+IPT and EAX for larger instances. We must also consider that LKH+GPX2 is a newly developed algorithm which may not have fully optimised implementation methods, and any refinements may significantly improve its future performance.

Acknowledgements. This work was supported by the Leverhulme Trust [award number RPG-2015-395] and by the UK's Engineering and Physical Sciences Research Council [grant number EP/J017515/1]. Results were obtained using the EPSRC-funded ARCHIE-WeSt High Performance Computer (www.archie-west.ac.uk, EPSRC grant EP/K000586/1).

Data Access. All data generated for this research are openly available from the Stirling Online Repository for Research Data (http://hdl.handle.net/11667/127).

References

1. Applegate, D.L., Bixby, R.E., Chvátal, V., Cook, W.J.: The Traveling Salesman Problem: A Computational Study. Princeton University Press, Princeton (2007)
2. Helsgaun, K.: Effective implementation of the Lin-Kernighan traveling salesman heuristic. Eur. J. Oper. Res. **126**(1), 106–130 (2000)
3. Helsgaun, K.: An effective implementation of k-opt moves for the Lin-Kernighan TSP heuristic. Technical report 109, Roskilde University (2007)
4. Nagata, Y., Kobayashi, S.: A powerful genetic algorithm using edge assembly crossover for the traveling salesman problem. INFORMS J. Comput. **25**(2), 346–363 (2013)

5. Whitley, D., Hains, D., Howe, A.: A hybrid genetic algorithm for the traveling salesman problem using generalized partition crossover. In: Schaefer, R., Cotta, C., Kołodziej, J., Rudolph, G. (eds.) PPSN 2010. LNCS, vol. 6238, pp. 566–575. Springer, Heidelberg (2010). https://doi.org/10.1007/978-3-642-15844-5_57

6. Corne, D.W., Reynolds, A.P.: Optimisation and generalisation: footprints in instance space. In: Schaefer, R., Cotta, C., Kołodziej, J., Rudolph, G. (eds.) PPSN 2010. LNCS, vol. 6238, pp. 22–31. Springer, Heidelberg (2010). https://doi.org/10.1007/978-3-642-15844-5_3

7. Smith-Miles, K., Lopes, L.: Measuring instance difficulty for combinatorial optimization problems. Comput. Oper. Res. **39**(5), 875–889 (2012)

8. Pihera, J., Musliu, N.: Application of machine learning to algorithm selection for TSP. In: IEEE 26th International Conference on Tools with Artificial Intelligence, pp. 47–54 (2014)

9. Kotthoff, L., Kerschke, P., Hoos, H., Trautmann, H.: Improving the state of the art in inexact TSP solving using per-instance algorithm selection. In: Dhaenens, C., Jourdan, L., Marmion, M.-E. (eds.) LION 2015. LNCS, vol. 8994, pp. 202–217. Springer, Cham (2015). https://doi.org/10.1007/978-3-319-19084-6_18

10. Kerschke, P., Kotthoff, L., Bossek, J., Hoos, H.H.: Leveraging TSP solver complementarity through machine learning. Evol. Comput. **26**(4), 597–620 (2018)

11. Bossek, J.: netgen: Network Generator for Combinatorial Graph Problems (2016). https://CRAN.R-project.org/package=netgen, R package version 1.3

12. R Core Team: R: A language and environment for statistical computing, Vienna (2018). https://www.R-project.org/

13. Mersmann, O., Bischl, B., Trautmann, H., Wagner, M., Bossek, J., Neumann, F.: A novel feature-based approach to characterize algorithm performance for the traveling salesperson problem. Ann. Math. Artif. Intell. **69**(2), 151–182 (2013)

14. Reinelt, G.: TSPLIB - a traveling salesman problem library. ORSA J. Comput. **3**(4), 376–384 (1991)

15. Rohe, A.: VLSI data sets (2017). http://www.math.uwaterloo.ca/tsp/vlsi/. Accessed 5 Nov 2018

16. Tinós, R., Helsgaun, K., Whitley, D.: Efficient recombination in the Lin-Kernighan-Helsgaun traveling salesman heuristic. In: Auger, A., Fonseca, C.M., Lourenço, N., Machado, P., Paquete, L., Whitley, D. (eds.) PPSN 2018. LNCS, vol. 11101, pp. 95–107. Springer, Cham (2018). https://doi.org/10.1007/978-3-319-99253-2_8

17. Lin, S., Kernighan, B.W.: An effective heuristic algorithm for the traveling-salesman problem. Oper. Res. **21**, 498–516 (1973)

18. Nagata, Y., Kobayashi, S.: Edge assembly crossover: a high-power genetic algorithm for the travelling salesman problem. In: ICGA, pp. 450–457 (1997)

19. Applegate, D., Bixby, R., Chvátal, V., Cook, W.: Concorde TSP solver (2003). http://www.math.uwaterloo.ca/tsp/concorde.html

20. Wolpert, D.H., Macready, W.G.: No free lunch theorems for optimization. IEEE Trans. Evol. Comput. **1**(1), 67–82 (1997)

21. Bischl, B., Mersmann, O., Trautmann, H., Preuß, M.: Algorithm selection based on exploratory landscape analysis and cost-sensitive learning, p. 313 (2012)

Runtime Analysis of Discrete Particle Swarm Optimization Applied to Shortest Paths Computation

Alexander Raß[✉], Jonas Schreiner, and Rolf Wanka

Department of Computer Science, University of Erlangen-Nuremberg, Erlangen, Germany
{alexander.rass,jonas.schreiner,rolf.wanka}@fau.de

Abstract. We mathematically analyze a discrete particle swarm optimization (PSO) algorithm solving the single-source shortest path (SSSP) problem. Key features are an improved and extended study on Markov chains expanding the adaptability of this technique and its application on the well-known SSSP problem. The results are upper and lower bounds on the expected optimization time. For upper bounds, we combine return times within a Markov model with the well known fitness level method which is appropriate even for the non-elitist PSO algorithm. For lower bounds we prove that the recently introduced property of indistinguishability applies in this setting and we also combine it with a further Markov chain analysis. We prove a cubic upper and a quadratic lower bound and an exponential upper and lower bound on the expected runtime, respectively, depending on a PSO parameter.

Keywords: Discrete particle swarm optimization · Runtime analysis · Single-source shortest paths · Markov chains

1 Introduction

In practice, optimization problems often do not have one single algorithm solving every problem instance in the best possible way. Instead there is a wide variety of algorithms each having situations where it works best in performance, quality or other measures. Meta-heuristics can be used successfully for optimizing a large number of problems, even if there is almost nothing known about the objective function which has to be optimized. That is, the objective function is given in a black-box setting.

However, due to a difficult structure of such algorithms and the applied problems, it is hard to prove performance indicators, i.e., quality guarantees, as well as estimating the number of steps until some criterion is met. In the present work we consider measuring the expected time it takes until a specific meta-heuristic, more precisely, the Particle Swarm Optimization (PSO) algorithm, finds the optimal solution.

A. Liefooghe and L. Paquete (Eds.): EvoCOP 2019, LNCS 11452, pp. 115–130, 2019.
https://doi.org/10.1007/978-3-030-16711-0_8

For reasons of comparison to specially designed algorithms, it is desired that good meta-heuristics solve black-box versions of tractable problems provably fast, i.e., in polynomial time. To prove this desired behavior, we use here the classical single-source shortest path problem as a reference for a discrete version of PSO. For a swarm of size one, we obtain a complete classification of the runtime for the single-source shortest path problem depending on the algorithm's parameter.

Runtime bounds are available for related nature-inspired meta-heuristics, for example evolutionary algorithms (EAs) [1,2] and ant colony optimization (ACO) [3,4], which are algorithms devoted to discrete domains. Another bio-inspired meta-heuristic originally devoted to continuous domains developed by Eberhart and Kennedy [5,6] is Particle Swarm Optimization (PSO). Schmitt and Wanka [7] proved that a slightly modified PSO finds local optima almost surely in the setting of continuous objective functions. PSO has already been adapted to multiple discrete problems, for instance for binary problems [8] and the traveling salesperson problem [9,10].

The PSO algorithm is inspired by the behavior of bird flocks or fish schools searching for good locations for residence. The algorithm organizes a collection (swarm) of particles, the bird flock in nature, where each of them has an admissible solution and a velocity. Additionally the best solution found by a single particle which is called local attractor is stored. Particles are interacting by sharing the global attractor, the best solution found by all particles. The particles move according to their velocity partially redirected by a randomly perturbated pull towards its local and global attractor. The algorithm's parameters balance the influence of the attractors to have a tradeoff between exploration and exploitation.

Several authors have adapted PSO to discrete domains which requires a complete reinterpretation of movement within the search space. We will focus on the OnePSO proposed in [11] because it can be adapted easily to various optimization problems.

The main contribution of this paper is the extension of the Markov model used in [11]. In [11], transitions are only allowed to neighboring states, whereas here also self-loops are allowed. Interestingly, the SSSP problem provides a showcase where such self-loops are actually needed. We will see that the computation of expected return times in this Markov model extended by self-loops is possible, and, hence, we provide an easy to use mathematical analysis tool for applications extending those from [11].

We will evaluate the optimization time in the number of function evaluations of the objective function as this is the established method for measuring the algorithm runtime.

Related Work. The results relevant to this work concern the $(1+1)$-EA algorithm and ACO algorithm. We give a brief overview of two related results for meta-heuristics on discrete problems.

The authors of [12] propose an ACO algorithm for the single destination shortest path (SDSP) problem as well as for the all-pairs shortest path (APSP) problem. Among other results they prove upper bounds of $\mathcal{O}(\Delta \cdot l \cdot l^* + l/\rho)$ expected iterations with n objective function evaluations in each iteration on SDSP and $\mathcal{O}(\Delta \cdot l \cdot l^*/\rho)$ expected iterations with n^2 objective function evaluations in each iteration on APSP, where Δ is the maximum degree, l is the maximum number of edges on any shortest path, $l^* = \max\{l, \ln n\}$ and ρ is the evaporation factor of the ACO algorithm.

In [1] one of the most comparable results can be found. There a $(1+1)$-EA performing optimization on the same genotype as we will do with the ONEPSO. The authors bound the expected optimization time of SSSP on n vertices by $\mathcal{O}(n^3)$ and $\Omega(n^2)$. We will receive the same bounds in the case if we choose the algorithm parameter $c > 1/2$, the probability of moving towards the attractor. In [2] the results are improved to $\mathcal{O}(n^2 \cdot l^*)$ with high probability and for special graphs they also have a lower bound of $\Omega(n^2 \cdot l^*)$ with high probability.

Organization of Paper. The paper is organized as follows. In Sect. 2, we present in general the discrete PSO algorithm that is investigated in this paper. The SSSP problem description and used data structures are described in Sect. 3. In Sect. 4, we describe the mathematical runtime model based on Markov chains with self-loops. Section 5 presents the technical basis on the Markov model necessary for the analysis of the PSO applied to the SSSP problem and Sect. 6 provides the Runtime analysis.

2 Discrete PSO

We now describe the ONEPSO algorithm introduced in [11], which minimizes discrete functions. Given is a finite set X, the search space, and an *objective function* $f : X \to \mathbb{R}$. The task is to find an optimal $x^* \in X$ that minimizes f. ONEPSO consists of a single particle that iteratively explores solution candidates to find such an optimal x^*. In the original PSO, a particle's next position is always within the current neighborhood. To apply ONEPSO to a problem one has to define a neighborhood relation on the search space. The set of neighbors of a node $x \in X$ is denoted by $\mathcal{N}(x) \subseteq X$. Interpreting solutions as nodes and neighborhoods as edges results in the *solution graph*.

Additionally we define a metric on the solution graph by the minimal number of edges on a path between two solutions.

Algorithm 1 describes the exact workings of ONEPSO. Input is the objective function $f : X \to \mathbb{R}$ and the parameter $c \in (0, 1]$ determining the probability of a move towards the best solution found so far, which is called the attractor a. ONEPSO starts by selecting an initial position $x \in X$ u.a.r. and updating the attractor a. After initialization, the steps of the particle are calculated iteratively. There are two possibilities, either the particle moves to a random neighbor or towards the attractor. With probability c the ONEPSO performs a move towards the attractor, i.e., a move which decreases the minimal distance to the attractor in the solution graph. Otherwise the particle moves to a random neighbor, which may also decrease the distance to the attractor. A movement towards the attractor is called an *attractor move*. If the function value of a new position is smaller, i.e., better, than the function value of the attractor, the attractor is updated.

Algorithm 1 ONEPSO

Require: $f : X \to \mathbb{R}, c \in]0, 1]$
 1: choose position $x \in X$ u.a.r. (uniformly at random)
 2: $a \leftarrow x$
 3: **loop**
 4:     choose $q \in]0, 1]$ u.a.r.
 5:     **if** $x = a$ **or** $q > c$ **then**
 6:         choose $x' \in \mathcal{N}(x)$ u.a.r.
 7:     **else**
 8:         choose $x' \in \{y \in \mathcal{N}(x) \mid \text{dist}_X(a, y) < \text{dist}_X(a, x)\}$ u.a.r.
 9:     **end if**
10:     $x \leftarrow x'$
11:     **if** $f(x) < f(a)$ **then**
12:         $a \leftarrow x$
13:     **end if**
14: **end loop**

---

3 Problem Structure

We now introduce the single-source shortest path problem (SSSP). In the SSSP, a directed graph $G = (V, E)$ with nodes $V = \{1, ..., n\}$, edges $E \subseteq V \times V$ and a weight function $\lambda : E \to \mathbb{R}^+$ is given. If $e \notin E$, we set $\lambda(e) = \infty$. All shortest paths between the source node s and all other nodes are to be found. W.l.o.g., $s = n$. A path from s to $i \in V$ is called *s-i-path*.

3.1 Shortest Path Trees

In order to apply the ONEPSO algorithm to the shortest path problem, we need a suitable representation of *candidate solutions*. Since the goal is to find $n-1$ shortest paths, a suitable representation *must* contain information on $n-1$ paths. Nodes on any (shortest) path have *exactly* one predecessor, which allows us to represent solution candidates by a *predecessor subgraph* [13, p. 475], a data structure that is widely used in the context of shortest path computation.

A genotype of a shortest path tree is obtained by storing each node's *predecessor* in its (shortest) path from s. That is, given a node i and the path from s to i, we store its predecessor $v_i \in \{1, \ldots, n\} \setminus \{i\}$. A node must not have itself as its predecessor. Therefore a shortest path tree can be represented as a vector of predecessors $(v_1, v_2, \ldots, v_{n-1})$, where the i-th entry is the predecessor of node i. Note that there is no predecessor of the source node $s = n$. There are vectors that do *not* represent valid shortest path trees, e.g., when no node has the predecessor s, or the graph contains cycles.

We use this vector representation as nodes for the solution graph, and say that two nodes are adjacent, iff they can be transformed into one another with a local operation. This operation can be chosen arbitrarily. We will use *transpositions*, which replace the predecessor of a node by a new one:

Definition 1 (Transpositions). *Let $(v_1, \ldots, v_{n-1})$ be a shortest path tree. A transposition is a tuple (i, j), with $i \in \{1, \ldots, n-1\}$ and $j \in \{1, \ldots, n\} \setminus \{i, v_i\}$. Applying (i, j) to a shortest path tree changes the predecessor v_i of the i-th node to j. In concrete terms: applying (i, j) to $(v_1, \ldots, v_i, \ldots, v_{n-1})$ results in the tree $(v_1, \ldots, j, \ldots, v_{n-1})$. Since there are $n - 1$ places, and each predecessor can be replaced by $n - 2$ others, the total number of transpositions is $(n - 1)(n - 2)$.*

3.2 Objective Function

Now we define the objective function we are going to use for the analysis.

Definition 2 (Objective Function). *Let $v = (v_1, \ldots, v_{n-1})$ be the representation of a predecessor subgraph (and $s = n$). We define the objective function*

$$f(v) = (\mathrm{dist}(v, s, 1), \mathrm{dist}(v, s, 2), \ldots, \mathrm{dist}(v, s, n - 1)),$$

where $\mathrm{dist}(v, s, i)$ is the distance from s to i in the predecessor subgraph defined by v, if there exists a valid path from s to i in that subgraph and ∞ otherwise.

To be able to compare function values we define a partial order. Let v, v' be solution candidates, then

$$f(v') < f(v) \Leftrightarrow \forall i : \mathrm{dist}(v', s, i) \leq \mathrm{dist}(v, s, i) \wedge \exists j : \mathrm{dist}(v', s, j) < \mathrm{dist}(v, s, j).$$

This objective function is often used if an EA is optimizing SSSP [1,2]. In [14], a single objective approach is analyzed in the context of a $(1 + 1)$-EA solving SSSP.

There is exactly one vector $\delta^* = (\delta_1^*, \ldots, \delta_{n-1}^*)$, which contains the lengths of *all shortest* paths from s. There may be multiple search points v, such that $f(v) = \delta^*$, but it is irrelevant which one is found.

4 Model

We will present a *general* way to model the behavior of the ONEPSO algorithm. Note that this model does not depend on a specific problem. For the upper runtime bounds we assume that the ONEPSO algorithm can find better positions exclusively at the attractor. This means that the runtime depends mainly on the number of steps ONEPSO needs to return to the attractor, if no improvement was achieved. We are going to model the behavior of ONEPSO *between two improvement attempts* as a Markov process.

If the particle is not at the attractor, there are three possibilities for the next move. Either the particle does an attractor move, moves away from the attractor, or keeps the current distance to the attractor. We use the *fitness level method* [15] and partition the search space into the levels $S_0, S_1, \ldots, S_n$. Please note that for SSSP we have only levels $S_0, \ldots, S_{n-1}$ as we store only $n-1$ predecessors (root has no predecessor) and therefore the diameter in X is only $n-1$. Assuming a is the attractor, then

$$S_i := \{x \in X \mid \text{dist}_X(a, x) = i\}.$$

If $x \in S_i$ is the current position, then ONEPSO can move to either S_{i-1}, S_{i+1}, or stay in S_i. ONEPSO does attractor moves with probability c and random moves with probability $1-c$. The fitness level after a random move depends on the number of neighbors and the distribution of the neighbors over S_{i-1}, S_i and S_{i+1}. The probabilities to move to S_{i-1}, stay in S_i and move to S_{i+1} are denoted by p_i^+, $p_i^=$ and p_i^-, respectively. For ONEPSO these probabilities are

$$p_i^+ := c + (1-c) \cdot |\mathcal{N}(x) \cap S_{i-1}| / |\mathcal{N}(x)|, \tag{1}$$

$$p_i^= := (1-c) \cdot |\mathcal{N}(x) \cap S_i| / |\mathcal{N}(x)|, \tag{2}$$

$$p_i^- := (1-c) \cdot |\mathcal{N}(x) \cap S_{i+1}| / |\mathcal{N}(x)|. \tag{3}$$

Note that state S_0 contains only the single solution a and therefore ONEPSO will certainly be in S_1 after one move. Conversely, in S_n the distance cannot be increased. This implies p_0^+, $p_0^=$ and p_n^- are zero and $p_0^- = 1$. The model is visualized in Fig. 1. An instance of this Markov model is denoted by $\mathcal{M}((p_i^+, p_i^=, p_i^-)_{1 \le i \le n})$.

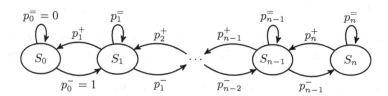

Fig. 1. State diagram of the Markov model with states $S_0, \ldots, S_n$ and corresponding state transition probabilities.

Also in [16] Markov chains are covered and equations similar to the following equations are obtained, which are used as a starting point for further analysis and insights.

We are interested in the expected number of steps to move from the state S_i to S_{i-1}. We call this number of steps H_i. Suppose ONEPSO is currently in S_i. With probability $p_i^=$ ONEPSO remains in S_i and then needs H_i steps to get to S_{i-1}. With probability p_i^- the particle moves to S_{i+1} and has to get back to S_i

before the next improvement attempt. Therefore the expected number of steps is $H_{i+1} + H_i$. In case of success ONEPSO needs just one step. Altogether we get

$$H_i = 1 + p_i^= \cdot H_i + p_i^- \cdot (H_i + H_{i+1}) \overset{1=p_i^+ +p_i^= +p_i^-}{\Leftrightarrow} H_i = \frac{1}{p_i^+} + \frac{p_i^-}{p_i^+} \cdot H_{i+1} \quad (4)$$

$$H_n = \frac{1}{p_n^+}. \quad (5)$$

We expand the expression for H_1 k times and get

$$H_1 = \sum_{i=1}^{k} \left(\frac{1}{p_i^+} \cdot \prod_{j=1}^{i-1} \frac{p_j^-}{p_j^+} \right) + H_{k+1} \cdot \prod_{i=1}^{k} \frac{p_i^-}{p_i^+}. \quad (6)$$

For $k = n - 1$ we get

$$H_1 = \sum_{i=1}^{n-1} \left(\frac{1}{p_i^+} \cdot \prod_{j=1}^{i-1} \frac{p_j^-}{p_j^+} \right) + H_n \cdot \prod_{i=1}^{n-1} \frac{p_i^-}{p_i^+}. \quad (7)$$

To receive an upper bound on H_i we set $p_i^- = 1 - p_i^+$ and omit $p_i^=$. Instances of this model are denoted by $\mathcal{M}((p_i^+)_{1 \le i \le n})$. Application of Eq. (4) in this simplified model leads to

$$H_i = \frac{1}{p_i^+} (1 + H_{i+1}) - H_{i+1} \qquad H_n = \frac{1}{p_n^+} \quad (8)$$

and its corresponding expansion results in

$$H_1 = \sum_{i=1}^{n} \left(\frac{1}{p_i^+} \cdot \prod_{j=1}^{i-1} \frac{1 - p_j^+}{p_j^+} \right) - \frac{1}{p_n^+} \prod_{j=1}^{n} \frac{1 - p_j^+}{p_j^+}. \quad (9)$$

5   Technical Results

If all probabilities p_i^+ are constant, Eq. (8) becomes a linear recurrence and can be solved with standard methods.

Theorem 1 ([11], Theorem 1). *Let $p_1^+ = p_2^+ = \ldots = p_{n-1}^+ = p$ be constant for some $p \in {]0, 1]}$ and $p_n^+ = 1$, then the expected return time H_1 to S_0 is*

$$H_1 = \begin{cases} \frac{1 - 2p\left(\frac{1-p}{p}\right)^n}{2p - 1} & \text{for } p \ne 1/2 \\ 2n - 1 & \text{otherwise}. \end{cases}$$

If p_i^+ is of the form $p_i^+ = 1/2 + i/2A(n)$, where $A(n) : \mathbb{N} \to \mathbb{N}$ is some non-decreasing function of n such that $\lim_{n \to \infty} A(n) = \infty$, we can approximate Expression (9).

Theorem 2 ([11], **Theorem 3**). *Let* $M = \mathcal{M}((p_i^+)_{1 \leq i \leq n})$, *with* $p_i^+ = 1/2 + i/2A(n)$, *then* $H_1 = \Theta(\min(\sqrt{A(n)}, n))$.

For the simplified model [17] presented the following theorem, which is proved by approximating the sum in Eq. (9) by an integral.

Theorem 3 ([17], **Theorem 7**). *Let* $M = \mathcal{M}((p^+(i))_{1 \leq i \leq n})$, *with* $p^+ : [0, n] \rightarrow]0, 1]$, p^+ *non-decreasing, then*

$$H_1 = \Omega\left(\text{base}(p^+, n)^n\right),$$
$$H_1 = \mathcal{O}\left(n \cdot \text{base}(p^+, n)^n\right) \text{ and}$$
$$H_1 = \Theta^*\left(\text{base}(p^+, n)^n\right), \text{where}$$

$$\text{base}(p^+, n) = \sup_{k \in [0, n[} \exp\left(\int_0^{\frac{k}{n}} \ln\left(\frac{1 - p^+(n \cdot x)}{p^+(n \cdot x)}\right) dx\right).$$

$\text{base}(p^+, n)$ *is maximized by the* k, *which satisfies* $p^+(k) = 1/2$.

This theorem can be extended easily to our full model.

Theorem 4. *Let* $M = \mathcal{M}((p^+(i), p^=(i), p^-(i))_{1 \leq i \leq n})$, *with* $p^+ : [0, n] \rightarrow]0, 1]$, $p^=, p^- : [0, n] \rightarrow [0, 1]$, $(p^-(i)/p^+(i))$ *is non-increasing, then*

$$H_1 = \Omega\left(\text{base}(p^+, p^-, n)^n\right),$$
$$H_1 = \mathcal{O}\left(n \cdot \text{base}(p^+, p^-, n)^n\right) \text{ and}$$
$$H_1 = \Theta^*\left(\text{base}(p^+, p^-, n)^n\right), \text{where}$$

$$\text{base}(p^+, p^-, n) = \sup_{k \in [0, n[} \exp\left(\int_0^{\frac{k}{n}} \ln\left(\frac{p^-(n \cdot x)}{p^+(n \cdot x)}\right) dx\right).$$

$\text{base}(p^+, p^-, n)$ *is maximized by the* k, *which satisfies* $p^+(k) = p^-(k)$.

Proof. One can easily adapt the proof presented in [17], by replacing the quotient $(1 - p(i))/p(i)$ with $p^-(i)/p^+(i)$. Note that to ensure that the quotient $p^-(i)/p^+(i)$ is non-increasing it suffices to have $p^+(i)$ non-decreasing and $p^-(i)$ non-increasing. $\blacksquare$

6   Runtime Analysis

We get an upper bound on the expected time it takes until ONEPSO finds an optimal solution by assuming the $n - 1$ predecessors are found iteratively and improvements are only achieved by moves from the attractor. For lower bounds we consider the expected time it takes to perform the very last improvement.

First of all, we need closed form expressions for the state transition probabilities of the Markov model. The probabilities can be calculated by Eqs. (1)–(3). This means we have to determine $\mathcal{N}(x)$ and S_i, S_{i-1} and S_{i+1} respectively, for given n, i and x. Let $x \in S_i$, then the set of neighbors $\mathcal{N}(x)$ is determined by all possible transpositions. There are $(n-1)(n-2)$ possible transitions according to Definition 1. The current level is the i-th, therefore the number of different entries among x and the attractor is exactly i. For each of these i entries, there exists exactly one transposition which replaces the entry with that of the attractor. The probability to move to S_{i-1} is thus

$$p_i^+ = c + (1-c) \cdot \frac{i}{(n-1)(n-2)}. \tag{10}$$

All other transpositions that change the predecessors of one of the i differing entries cause ONEPSO to remain in S_i. Thus the probability to stay in S_i is

$$p_i^= = (1-c) \cdot \frac{i(n-3)}{(n-1)(n-2)}. \tag{11}$$

Therefore $p_i^=$ is usually not zero and the reduced model without self loops is not sufficient. Since $p_i^+ + p_i^= + p_i^- = 1$ we can directly determine the probability to move to S_{i+1}.

$$p_i^- = 1 - p_i^+ - p_i^= = (1-c) \cdot \left(1 - \frac{i}{n-1}\right) \tag{12}$$

Please note that these values are independent of $x \in S_i$.

Now we will calculate H_1 for different values of c.

Lemma 1. *Let $c = 1/2$ and $M = \mathcal{M}((p_i^+, p_i^=, p_i^-)_{1 \le i \le n})$, with p_i^+, $p_i^=$ and p_i^- as specified in Equations (10)–(12), then $H_1 = \Theta(\sqrt{n})$.*

Proof. Let $c = 1/2$. For a lower bound we increase the probability p_i^+ of moving towards the attractor by $p_i^=$ and omit the chance to stay in S_i:

$$\tilde{p}_i^+ := p_i^+ + p_i^= = \frac{1}{2} + \frac{i}{2(n-1)}.$$

Now we can apply Theorem 2 to $\mathcal{M}((\tilde{p}_i^+)_{1 \le i < n})$, which results in $H_1 = \Omega(\sqrt{n})$.

For an upper bound we scale p_i^+ and p_i^- with $1/(1-p_i^=)$ and omit the probability $p_i^=$ from the model. This again allows us to use the simpler Markov model with two state transition possibilities $\mathcal{M}\left((\hat{p}_i^+)_{1 \le i \le n-1}\right)$ with $\hat{p}_i^+ = p_i^+/(1-p_i^=)$. This change decreases the expected optimization time by the factor $1/(1-p_i^=)$, the expected number of iterations ONEPSO can stay in the same state, but this value is at most 2 as $c = 1/2$ and $1/2 \le 1 - p_i^= \le 1$. Therefore the expected optimization time is changed only by a constant factor which can be neglected.

$$\hat{p}_i^+ = \frac{\frac{1}{2} + \frac{i}{2(n-1)(n-2)}}{1 - \frac{i(n-3)}{2(n-1)(n-2)}} \ge \frac{\frac{1}{2}}{\frac{2(n-1)(n-2)-i(n-3)}{2(n-1)(n-2)}}$$

$$= \frac{1}{2}\left(1 + \frac{i(n-3)}{2(n-1)(n-2)-i(n-3)}\right) \overset{n \ge 4}{\ge} \frac{1}{2}\left(1 + \frac{i}{4(n-1)}\right) =: \hat{p}_i^+$$

Now we can apply Theorem 2 to $\mathcal{M}((\hat{p}_i^+)_{1 \le i \le n-1})$ to receive the upper bound $H_1 = \mathcal{O}(\sqrt{n})$, which completes the proof.

Lemma 2. *Let $c \in]0, 1/2[$ and $M = \mathcal{M}((p_i^+, p_i^=, p_i^-)_{1 \leq i \leq n})$, with p_i^+, $p_i^=$ and p_i^- as specified in Eqs. (10)–(12) then for any arbitrarily small $\varepsilon > 0$ the expected return time to the attractor is bounded by*

$$H_1 = \mathcal{O}\left(n \cdot \alpha(c)^n\right) \text{ and } H_1 = \Omega\left((\alpha(c) - \varepsilon)^n\right), \text{where}$$
$$\alpha(c) = e^{-(1-2c)/(1-c)} \cdot \left(\tfrac{1-c}{c}\right).$$

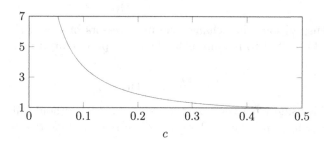

Fig. 2. The function $\alpha(c)$ with $c \in]0, 1/2]$

Proof. To apply Theorem 4 we have to convert the discrete probabilities into continuous functions. We achieve this by extending the domains of Eqs. (10)–(12) to $\mathbb{R}$. This is allowed, since these equations can be applied to any $i \in \mathbb{R}$. As p^+ is non-decreasing and p^- is non-increasing the fraction p^-/p^+ is non-increasing. The k, which solves $p^+(k) = p^-(k)$, maximizes the integral.

$$p^+(k) \stackrel{!}{=} p^-(k) \Leftrightarrow k = (n-2) \cdot (1 - 2c)/(1 - c)$$

Now we can calculate a value for $\text{base}(p^+, p^-, n-1)$ as specified in Theorem 4:

$$\text{base}(p^+, p^-, n-1) = \exp\left(\int_0^{\frac{k}{n-1}} \ln\left(\frac{p^-((n-1)\cdot x)}{p^+((n-1)\cdot x)}\right) \, dx\right)$$

$$= \exp\left(\int_0^{\frac{n-2}{n-1} \cdot \frac{1-2c}{1-c}} \ln\left(\frac{(1-c)(1-x)}{x + (1-c) \cdot \frac{x}{n-2}}\right) \, dx\right)$$

The indefinite integral can be calculated by

$$\int \ln\left(\frac{(1-c)(1-x)}{c + (1-c) \cdot \frac{x}{n-2}}\right) \, dx$$

$$= \int \ln(1-c) \, dx + \int \ln(1-x) \, dx - \int \ln\left(c + (1-c) \cdot \frac{x}{n-2}\right) \, dx$$

$$= [x \cdot \ln(1-c)] + [-x - (1-x) \cdot \ln(1-x)]$$

$$+ \left[-\ln\left(c + (1-c) \cdot \frac{x}{n-2}\right) \cdot \left(x + \frac{c \cdot (n-2)}{1-c}\right) + x\right]$$

$$= \ln \left(\frac{(1-c)(1-x)}{c + \frac{(1-c)}{n-2} \cdot x} \right) \cdot x - \ln(1-x) - \frac{c \cdot (n-2)}{1-c} \cdot \ln \left(c + \frac{1-c}{n-2} \cdot x \right) =: I(x).$$

We get the definite integral by evaluating the bounds.

$$I \left(\frac{n-2}{n-1} \cdot \frac{1-2c}{1-c} \right) - I(0)$$

$$= \left[0 - \ln \left(\frac{c}{1-c} \cdot \left(1 + \frac{1}{n-1} \cdot \frac{1-2c}{c} \right) \right) - \frac{c \cdot (n-2)}{1-c} \cdot \ln \left(c + \frac{1-2c}{n-1} \right) \right]$$

$$- \left[0 - 0 - \frac{c \cdot (n-2)}{1-c} \cdot \ln(c) \right]$$

$$= - \ln \left(\frac{c}{1-c} \right) - \ln \left(1 + \frac{1}{n-1} \cdot \frac{1-2c}{c} \right) - \frac{c \cdot (n-2)}{1-c} \cdot \ln \left(1 + \frac{1-2c}{c \cdot (n-1)} \right)$$

$$= - \ln \left(\frac{c}{1-c} \right) - \frac{1}{n-1} \cdot \frac{1-2c}{c} + \mathcal{O}\left(n^{-2} \right)$$

$$- \frac{c \cdot (n-2)}{1-c} \cdot \left(\frac{1-2c}{c \cdot (n-1)} - \frac{1}{2} \cdot \left(\frac{1-2c}{c \cdot (n-1)} \right)^2 + \mathcal{O}\left(n^{-3} \right) \right)$$

$$= - \ln \left(\frac{c}{1-c} \right) - \frac{1}{n-1} \cdot \frac{1-2c}{c} + \mathcal{O}\left(n^{-2} \right)$$

$$- \frac{1-2c}{1-c} \cdot \left(1 - \frac{1}{n-1} \right) + \frac{(1-2c)^2}{(1-c) \cdot 2c} \cdot \frac{1}{n-1} - \mathcal{O}\left(n^{-2} \right)$$

$$= \ln \left(\frac{1-c}{c} \right) - \frac{1-2c}{1-c} - \frac{1}{n-1} \cdot \left(\frac{1-2c}{c} - \frac{1-2c}{1-c} - \frac{(1-2c)^2}{(1-c) \cdot 2c} \right) \pm \mathcal{O}\left(n^{-2} \right)$$

$$= \ln \left(\frac{1-c}{c} \right) - \frac{1-2c}{1-c} - \frac{1}{n-1} \cdot \frac{(1-2c)^2}{(1-c) \cdot 2c} \pm \mathcal{O}\left(n^{-2} \right).$$

Therefore we have

$$\text{base}(p^+, p^-, n-1) = \left(\frac{1-c}{c} \right) \cdot e^{-\frac{1-2c}{1-c}} \cdot e^{-\frac{1}{n-1} \cdot \frac{(1-2c)^2}{(1-c) \cdot 2c} \pm \mathcal{O}(n^{-2})}.$$

Using Theorem 4 concludes the proof, because $e^{-\frac{1}{n-1} \cdot \frac{(1-2c)^2}{(1-c) \cdot 2c} \pm \mathcal{O}(n^{-2})} \in [1 - \varepsilon, 1]$ for arbitrary small but positive value ε if n is large enough.

6.1 Upper Bounds

We use the fitness level method [15] and the Markov model to calculate the expected time ONEPSO needs to compute the shortest path tree *beginning with the root*. This means, that we have to find all shortest s-i-paths consisting of k edges, before we can find a shortest s-i-path consisting of $l > k$ edges. To achieve provable bounds the objective functions must be unimodal in the sense that a function $f : X \to Y$ is unimodal iff there is always a neighbor with a better function value, except at a global minimum. The multi-objective function described in Sect. 3.2 is unimodal according to this definition, since the current position v is either minimal, or there exists a position i, whose predecessor can be replaced by the optimal predecessor. Building the shortest path tree from the root provides an upper bound, since the tree can have some randomly correct positions upon initialization. Furthermore, we assume that better positions can *exclusively* be found at the attractor. Additionally we take advantage of the fact, that the attractor does not get updated, if *already found* shortest s-i-paths would be forgotten.

First some definitions to talk about edge-shortest paths.

Definition 3 ([1]). *Consider all shortest paths from s to i. By l_i we denote the minimal number of edges of all such paths. Furthermore,*

1. *let $m_t = |\{i \mid l_i = t\}|$ be the number of nodes, whose shortest s-i-path has minimal number of edges t*
2. *and let $l = \max\{l_i \mid i \in V\}$.*

Now we can compute the desired upper bounds.

Theorem 5. *The expected optimization time $T(n)$, to solve the single-source shortest path problem with n nodes is bounded from above by*

$$
T(n) = \begin{cases} \mathcal{O}(n^3) & if c \in]1/2, 1] \\ \mathcal{O}(n^{7/2}) & if c = 1/2 \\ \mathcal{O}\left(n^4 \cdot \alpha(c)^n\right) & if c \in]0, 1/2[. \end{cases}
$$

Proof. We consider a prototypical step of ONEPSO. Suppose the attractor has just been updated, then a t exists, with $0 \le t < n-1$, and all shortest paths to nodes i, with $l_i < t$, have been found. We calculate the expected time, ONEPSO needs to find the shortest paths to all nodes j, with $l_j = t$. A predecessor node u, on a shortest path to one of the nodes j, is a node with $l_u = t-1$ and the shortest path to u has already been found. The shortest path to a fixed node j would be found, if the transposition (j, u) is chosen. Since there are $(n-1)(n-2)$ transpositions and exactly one is desired, the probability is $1/(n-1)(n-2)$. We multiply this ratio with m_t and get $m_t/(n-1)(n-2)$ for the probability to find the first new shortest path. The inverse of this is the expected number of attempts for an improvement. If ONEPSO makes a move that does not improve the position, then the particle has to return to the attractor, which takes H_1 steps. For a fixed t this results in an expected optimization time of up to

$$
\sum_{i=1}^{m_t} \left((H_1 + 1) \left(\frac{n^2}{i} - 1 \right) + 1 \right).
$$

Since t can have values $1, \ldots, l$, we get a total expected optimization time of

$$
\sum_{t=1}^{l} \sum_{i=1}^{m_t} \left((H_1 + 1) \left(\frac{n^2}{i} - 1 \right) + 1 \right) \le (H_1 + 1) \cdot n^2 \cdot \sum_{t=1}^{l} (\ln(m_t) + 1)
$$

$$= \mathcal{O}\left(H_1 \cdot n^3\right).$$

Using Chernoff bounds similarly as in [2] one can improve the upper bound and prove that $\mathcal{O}(H_1 \cdot n^2 \cdot \max\{l, \log(n)\})$ is an upper bound with high probability. As l can be as large as n we omit the maximum and use n instead. Now we calculate H_1 for different $c \in {]}0, 1]$. With $c \in {]}1/2, 1]$ the probability of moving towards the attractor is at least $c > 1/2$. Therefore $\mathcal{M}\left((c)_{1 \leq i \leq n-1}\right)$ supplies by Theorem 1 an upper bound for H_1 which is $\mathcal{O}(1)$. It follows, that $T(n) = \mathcal{O}(n^3)$ for $c > 1/2$. Let $c = 1/2$. We apply Lemma 1 and get $H_1 = \mathcal{O}(\sqrt{n})$ and thus $T(n) = \mathcal{O}(n^{7/2})$. Finally let $c \in {]}0, 1/2[$. We apply Lemma 2 and get $H_1 = \mathcal{O}\left(n \cdot \alpha(c)^n\right)$. With that we have $T(n) = \mathcal{O}\left(n^4 \cdot \alpha(c)^n\right)$.

6.2   Lower Bounds

We consider only the expected optimization time it takes for ONEPSO to successfully perform the *last* improvement, because the optimization time is dominated by this step. For this purpose we combine the property of indistinguishability introduced in [11] with lower bounds calculated for H_1 earlier in this paper. We repeat the concept of *indistinguishable states* to characterize the set of neighbors of the attractor.

Definition 4 (Definition 17, Indistinguishable States [11]). *Let M be a Markov process with a finite set of states Y and let $\overline{Y} \subseteq Y$. Furthermore, let $(Z_i)_{i \geq 0}$ be the sequence of visited states of M and let $T = \min\{t > 0 \mid Z_t \in \overline{Y}\}$. Then $\overline{Y}$ is called* indistinguishable *with respect to M if*

1. the initial state Z_0 is uniformly distributed over $\overline{Y}$, i.e., for all $y \in Y$:

$$P[Z_0 = y] = \mathbb{1}_{y \in \overline{Y}}/|\overline{Y}| := \begin{cases} 1/\overline{Y} & if y \in \overline{Y} \\ 0 & if y \notin \overline{Y} \end{cases}$$

2. and the probabilities to reach states in $\overline{Y}$ from states in $\overline{Y}$ are symmetric, i.e., for all $y_1, y_2 \in \overline{Y}$:

$$P[Z_T = y_2 | Z_0 = y_1] = P[Z_T = y_1 | Z_0 = y_2].$$

One may think of indistinguishable states as a plateau. But actually each of the indistinguishable states may have a different function value and indeed in our case at least one of these states, the optimum, is better than all other states in the indistinguishable set. Definition 4 is used to derive a lower limit for the expected number of steps needed, to visit a fixed state of the amount $\overline{Y}$.

Theorem 6 (Theorem 18, [11]). *Let M be a Markov process as in Definition 4 and let $\overline{Y}$ be indistinguishable with respect to M. Let $h(M) \in \mathbb{R}^+$, such that $E[T] \geq h(M)$, then the expected time to reach a fixed $y \in \overline{Y}$, is bounded below by $h(M) \cdot \Omega\left(|\overline{Y}|\right)$.*

We show that the neighbors of the attractor in the last step of ONEPSO are indistinguishable as in Definition 4, and estimate T, to get a lower limit on the optimization time as one of these indistinguishable states is the global optimum.

Lemma 3. *Let $\overline{Y} = X_1$ be the set of neighbor nodes in the last step of ONEPSO, and assume, that ONEPSO is at the attractor, then $\overline{Y}$ is indistinguishable as in Definition 4.*

Proof. 1. Since the particle is at the attractor, the particle moves to a random element in $X_1 = \overline{Y}$ with the next move. Since transpositions are chosen u.a.r. all states $Z_0 \in \overline{Y}$ have equal probability.

2. In $\overline{Y}$ there are only solutions that differ from the attractor a by a single value within the predecessor vector. If we inspect two solutions $u, w \in \overline{Y}$ then they either differ in the same position from the attractor or they differ in two different positions. Paths from u to w can be converted bijectively to paths from w to u which have the same probability by substituting predecessors in solutions appearing in the path. The substitutions on an intermediate solution v is performed as described in the following: First, we consider the case where u and w differ in the same index d compared to the attractor. If $v_d \in \{u_d, w_d\}$ then we change v_d to the other non equal value. Second, we consider the case where u and w differ not in the same index from the attractor a. Let d_u and d_w be the differing index in u and w compared to a respectively. A change in the index d_u in v has to be converted to a change in the index d_w, such that if $v_{d_u} = u_{d_u}$ or $v_{d_u} = a_{d_u}$ before substitution then $v_{d_w} = w_{d_w}$ or $v_{d_w} = a_{d_w}$ after substitution respectively.

As any path has a corresponding path in the opposite direction with the same probability we receive that the probabilities to reach states in $\overline{Y}$ from states in $\overline{Y}$ are symmetric according to Definition 4.

Theorem 7 *The expected optimization time $T(n)$ to solve the single-source shortest path problem with n nodes is bounded from below by*

$$
T(n) = \begin{cases}
\Omega(n^2) & if\ c \in \,]1/2, 1] \\
\Omega(n^{5/2}) & if\ c = 1/2 \\
\Omega\left((\alpha(c) - \varepsilon)^n\right) & if\ c \in \,]0, 1/2[.
\end{cases}
$$

Proof. We assume that the attractor has just been updated, and the only better position, the global optimum x^*, has distance one to the attractor, therefore $x^* \in X_1$. The set $\overline{Y} = X_1$ is indistinguishable, according to Lemma 3. Invoking Theorem 6 we get the lower bound $h(M) \cdot \Omega(|\overline{Y}|)$. To get a value for $h(M)$ we calculate $E[T]$, the expected time it takes until ONEPSO returns to $\overline{Y}$ when it is currently in a random position in $\overline{Y}$. If ONEPSO's current position is in $\overline{Y}$, then either ONEPSO can do an attractor move and returns to $\overline{Y}$ in the next step, or it can stay in $\overline{Y}$, or it can move to X_2 and then returns to $\overline{Y}$ with H_2 steps.

$$
E[T] = 2 \cdot p_1^+ + p_1^= + p_1^- \cdot (1 + H_2) = \Omega(H_2) =: h(M)
$$

We can bound H_2 by using the Recurrence (4).

$$H_1 = \frac{1}{p_1^+} + \frac{p_1^-}{p_1^+} \cdot H_2 \Leftrightarrow H_2 = \frac{1}{p_1^-} \cdot (p_1^+ \cdot H_1 - 1) \geq \frac{c \cdot H_1 - 1}{1 - c}.$$

It follows that $H_2 = \Omega(H_1)$, and thus the lower bound on the expected optimization time is $\Omega(H_1 \cdot |\overline{Y}|)$. For $c \in]0, 1/2[$ we invoke Lemma 2 and get $H_1 = \Omega((\alpha(c) - \varepsilon)^n)$. For $c = 1/2$ we apply Lemma 1 and get $H_1 = \Omega(\sqrt{n})$. Finally, for $c \in]1/2, 1]$ we choose the trivial bound $H_1 = \Omega(1)$.

7   Conclusion

We analyzed the expected optimization time needed for ONEPSO to optimize the SSSP. For this purpose we calculated expected return times on an extension to recently introduced Markov chains. Depending on the algorithm parameter c, which is the probability of moving towards the attractor, the expected optimization time may be polynomial ($c \geq 1/2$) and exponential ($c < 1/2$), respectively. Hereby we provide an easy to use mathematical analysis tool to evaluate Markov chains extending [11].

For future work one can apply the presented tools on further optimization problems. Also further algorithms could be analyzed. In particular, our tools may be useful in the analysis of non-elitist meta-heuristics, for instance the Strong Selection Weak Mutation (SSWM) evolutionary regime [18] as an example of a non-elitist algorithm.

Acknowledgement. The authors would like to thank Bernd Bassimir for useful discussions.

References

1. Scharnow, J., Tinnefeld, K., Wegener, I.: The analysis of evolutionary algorithms on sorting and shortest paths problems. J. Math. Model. Algorithms **3**(4), 349–366 (2004). https://doi.org/10.1023/B:JMMA.0000049379.14872.f5
2. Doerr, B., Happ, E., Klein, C.: Tight analysis of the (1+1)-EA for the single source shortest path problem. Evol. Comput. **19**(4), 673–691 (2011). https://doi.org/10.1162/EVCO_a_00047
3. Neumann, F., Witt, C.: Runtime analysis of a simple ant colony optimization algorithm. Algorithmica **54**(2), 243–255 (2007). https://doi.org/10.1007/s00453-007-9134-2
4. Doerr, B., Neumann, F., Sudholt, D., Witt, C.: On the runtime analysis of the 1-ANT ACO algorithm. In: Proceedings of the 9th ACM Genetic and Evolutionary Computation Conference (GECCO), pp. 33–40 (2007). https://doi.org/10.1145/1276958.1276964
5. Eberhart, R.C., Kennedy, J.: A new optimizer using particle swarm theory. In: Proceedings of the 6th International Symposium on Micro Machine and Human Science, pp. 39–43 (1995). https://doi.org/10.1109/MHS.1995.494215

6. Kennedy, J., Eberhart, R.C.: Particle swarm optimization. In: Proceedings of the IEEE International Conference on Neural Networks, vol. 4, pp. 1942–1948 (1995). https://doi.org/10.1109/ICNN.1995.488968

7. Schmitt, M., Wanka, R.: Particle swarm optimization almost surely finds local optima. Theoret. Comput. Sci. **561A**, 57–72 (2015). https://doi.org/10.1016/j.tcs.2014.05.017

8. Sudholt, D., Witt, C.: Runtime analysis of a binary particle swarm optimizer. Theoret. Comput. Sci. **411**(21), 2084–2100 (2010). https://doi.org/10.1016/j.tcs.2010.03.002

9. Clerc, M.: Discrete particle swarm optimization, illustrated by the traveling salesman problem. In: New Optimization Techniques in Engineering, vol. 141, pp. 219–239. Springer, Heidelberg (2004). https://doi.org/10.1007/978-3-540-39930-8_8

10. Hoffmann, M., Mühlenthaler, M., Helwig, S., Wanka, R.: Discrete particle swarm optimization for TSP: theoretical results and experimental evaluations. In: Bouchachia, A. (ed.) ICAIS 2011. LNCS (LNAI), vol. 6943, pp. 416–427. Springer, Heidelberg (2011). https://doi.org/10.1007/978-3-642-23857-4_40

11. Mühlenthaler, M., Raß, A., Schmitt, M., Siegling, A., Wanka, R.: Runtime analysis of a discrete particle swarm optimization algorithm on sorting and OneMax. In: Proceedings of the 14th ACM/SIGEVO Workshop on Foundations of Genetic Algorithms (FOGA), pp. 13–24 (2017). https://doi.org/10.1145/3040718.3040721

12. Sudholt, D., Thyssen, C.: Running time analysis of ant colony optimization for shortest path problems. J. Discrete Algorithms **10**, 165–180 (2012). https://doi.org/10.1016/j.jda.2011.06.002

13. Cormen, T.H., Leiserson, C.E., Rivest, R.L.: Introduction to Algorithms. MIT Press, McGraw-Hill, Cambridge (1990)

14. Baswana, S., Biswas, S., Doerr, B., Friedrich, T., Kurur, P.P., Neumann, F.: Computing single source shortest paths using single-objective fitness. In: Proceedings of the 10th ACM/SIGEVO Workshop on Foundations of Genetic Algorithms (FOGA), pp. 59–66 (2009). https://doi.org/10.1145/1527125.1527134

15. Wegener, I.: Methods for the analysis of evolutionary algorithms on pseudo-Boolean functions. In: Sarker, R., et al. (eds.) Evolutionary Optimization, pp. 349–369. Springer, Boston (2002). https://doi.org/10.1007/0-306-48041-7_14

16. Droste, S., Jansen, T., Wegener, I.: Dynamic parameter control in simple evolutionary algorithms. In: Proceedings of the 6th ACM/SIGEVO Workshop on Foundations of Genetic Algorithms (FOGA), pp. 275–294 (2001). https://doi.org/10.1016/B978-155860734-7/50098-6

17. Mühlenthaler, M., Raß, A., Schmitt, M., Wanka, R.: Exact Markov chain-based runtime analysis of a discrete particle swarm optimization algorithm on sorting and OneMax (2019). https://arxiv.org/abs/1902.01810, extended version of [11]

18. Gillespie, J.H.: Some properties of finite populations experiencing strong selection and weak mutation. Am. Nat. **121**(5), 691–708 (1983). https://doi.org/10.1086/284095

Quasi-Optimal Recombination Operator

Francisco Chicano[1]([✉])(iD), Gabriela Ochoa[2](iD), Darrell Whitley[3],
and Renato Tinós[4](iD)

[1] University of Malaga, Málaga, Spain
chicano@lcc.uma.es
[2] University of Stirling, Stirling, UK
gabriela.ochoa@cs.stir.ac.uk
[3] Colorado State University, Fort Collins, USA
whitley@cs.colostate.edu
[4] University of Sao Paulo, São Paulo, Brazil
rtinos@ffclrp.usp.br

Abstract. The output of an optimal recombination operator for two
parent solutions is a solution with the best possible value for the objective
function among all the solutions fulfilling the gene transmission property:
the value of any variable in the offspring must be inherited from one of the
parents. This set of solutions coincides with the largest dynastic potential
for the two parent solutions of any recombination operator with the gene
transmission property. In general, exploring the full dynastic potential is
computationally costly, but if the variables of the objective function have
a low number of non-linear interactions among them, the exploration can
be done in $O(4^{\beta}(n + m) + n^2)$ time, for problems with n variables, m
subfunctions and β a constant. In this paper, we propose a quasi-optimal
recombination operator, called Dynastic Potential Crossover (DPX), that
runs in $O(4^{\beta}(n+m)+n^2)$ time in any case and is able to explore the full
dynastic potential for low-epistasis combinatorial problems. We compare
this operator, both theoretically and experimentally, with two recently
defined efficient recombination operators: Partition Crossover (PX) and
Articulation Points Partition Crossover (APX). The empirical comparison uses NKQ Landscapes and MAX-SAT instances.

Keywords: Recombination operator · Dynastic potential ·
Gray box optimization

1 Introduction

Many binary recombination operators for genetic algorithms have the property
of *gene transmission* [1]. When the solutions are represented by a set of variables taking values from a set (possibly different for each of them) with no
other constraint among the variables, this property implies that any variable

This research is funded by the Spanish Ministry of Economy and Competitiveness and
FEDER under contract TIN2017-88213-R, and the University of Malaga.

© Springer Nature Switzerland AG 2019
A. Liefooghe and L. Paquete (Eds.): EvoCOP 2019, LNCS 11452, pp. 131–146, 2019.
https://doi.org/10.1007/978-3-030-16711-0_9

in any child will take the value of the same variable in one of the parents. In particular, the variables having the same value for both parents will have the same value in all the children (i.e., the *respect* property [1] is obeyed). The other (differing) variables will take one of the values coming from a parent solution. The set of all the solutions that can be generated by a recombination operator from two parents is called *dynastic potential*. If we denote by $d(x, y)$ the Hamming distance (number of differing variables) between two solutions x and y, the largest dynastic potential of a recombination operator is $2^{d(x,y)}$. Uniform crossover has this dynastic potential. The dynastic potential of single-point crossover has size $2d(x, y)$ and the one of two-point crossover has size $2d(x, y) + \binom{d(x,y)-1}{2} = 1 + d(x, y)(d(x, y) + 1)/2$. In general, z-point crossover has a dynastic potential of size $O(d(x, y)^z)$ for $z << n$, with n variables.

Our goal in this paper is to design an *Optimal Recombination Operator* [2], which is one obtaining the best offspring from the largest dynastic potential. In the worst case, however, such a recombination operator is computationally expensive, since finding the best offspring in the largest dynastic potential is an NP-hard problem. For this reason, we design a *Quasi-Optimal Recombination Operator*, with worst time complexity $O(4^\beta(n + m) + n^2)$ where m is the number of subfunctions and β is an arbitrary constant. This operator will find the best offspring of the largest dynastic potential if the objective function has low espitasis, that is, if the number non-linear interactions among variables is small.

Our proposed operator, called Dynastic Potential Crossover (DPX), uses the variable interaction graph of the objective function to simplify the evaluation of the $2^{d(x,y)}$ solutions in the dynastic potential by using dynamic programming. The ideas for this efficient evaluation date back to Hammer's basic algorithm for variable elimination [3] and are also commonly used in operations over Bayesian networks [4]. Since it requires more information than just the objective function to do the job, this operator is framed in the so-called gray box optimization [5].

Recently defined crossover operators similar to ours are Partition Crossover (PX) [6] and Articulation Points Partition Crossover (APX) [7]. Although they were proposed to work with pseudo-Boolean functions, they can also be applied to the more general representation of variables defined over a finite alphabet. PX and APX also use the variable interaction graph of the objective function to efficiently compute a good offspring among a large number of them. PX and APX have $O(n^2 + m)$ time complexity and both of them obtained excellent performance in different problems [7–10]. When combined with other gray box optimization operators, partition crossover was capable of optimizing instances with 1 million variables in seconds. We compare DPX with these two operators from a theoretical point of view and in the experimental section.

The paper is organized as follows. Section 2 presents the required background to understand the working principles of DPX. The proposed recombination operator is presented in Sect. 3. Section 4 describes the experiments and presents the results and, finally, Sect. 5 concludes the paper.

2   Background

We will work along the paper with functions defined over a set of variables x_i, each one taking values in a finite set, X_i, not necessarily the same for all the variables. We say that a function f of n variables has k-bounded epistasis if it can be written as a sum of m subfunctions f_l, each one depending on at most k variables:

$$f(x) = \sum_{l=1}^{m} f_l(x_{i_{l,1}}, x_{i_{l,2}}, \ldots, x_{i_{l,k}}), \tag{1}$$

where $i_{l,j}$ is the index of the j-th variable in subfunction f_l. In the case of binary variables, these functions have been named Mk Landscapes by Whitley et al. [5]. In *Gray Box Optimization*, the optimizer can evaluate the set of m subfunctions in Eq. (1) (although their internal structure is unknown). This contrasts with Black Box Optimization, where the optimizer can only evaluate solutions and get their fitness value.

2.1   Variable Interaction Graph

The *Variable Interaction Graph* (VIG) [5] is a useful tool that can be constructed under Gray Box Optimization. It is a graph $VIG = (V, E)$, where V is the set of variables and E is the set of edges representing all pairs of variables (x_i, x_j) having *nonlinear interactions*. These nonlinear interactions can be captured in two ways. First, assuming that every pair of variables appearing together in a subfunction have a nonlinear interaction. A second approach is to apply the Fourier transform [11], and then look at every pair of variables to determine if there is a non-zero Fourier coefficient associated to a term with the two variables. This second method is more precise and not very expensive, because the Fourier transform can be constructed in $O(n)$ time for k-bounded epistasis functions.

An example of the construction of the variable interaction graph for a function with $n = 18$ variables (numbered from 0 to 17) and $k = 3$, is given below. We will refer to variables using numbers, e.g., $9 = x_9$. The objective function is the sum over the following 18 subfunctions:

$$
\begin{array}{llll}
f_0(0,6,14) & f_5(5,4,2) & f_{10}(10,2,17) & f_{15}(15,7,13) \\
f_1(1,0,6) & f_6(6,10,13) & f_{11}(11,16,17) & f_{16}(16,9,11) \\
f_2(2,1,6) & f_7(7,12,15) & f_{12}(12,10,17) & f_{17}(17,5,2) \\
f_3(3,7,13) & f_8(8,3,6) & f_{13}(13,12,15) & \\
f_4(4,1,14) & f_9(9,11,14) & f_{14}(14,4,16) &
\end{array}
$$

From these subfunctions, assume we extract the nonlinear interactions that are shown in Fig. 1. In this example, every pair of variables that appear together in a subfunction has a nonlinear interaction.

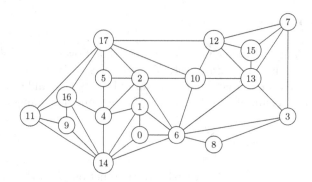

Fig. 1. Sample Variable Interaction Graph (VIG).

2.2 Recombination Graph

Let us assume that we have two solutions to recombine using the optimal recombination operator. We call these two solutions the *red* and the *blue* parents. All the variables with the same value in both parents will also share the same value in the offspring and the solutions in the dynastic potential will be in a hyperplane determined by the common variables. In the solution representation we will use digit 0 to denote that the variable has the same value as in the red parent and 1 to denote that the value is different. Thus, the red solution will be the string with all 0s. For example, let the two parents be

$$red = 000000000000000000 \quad \text{and} \quad blue = 111101011101110110$$

in our sample function of Sect. 2.1. Therefore, x_4, x_6, x_{10}, x_{14}, and x_{17} are identical in both parents. The rest of the variables are different. Both parents reside in a hyperplane denoted by $h = ****0*0***0***0**0$ where $*$ denotes the variables that are different in the two solutions, and 0 marks the positions where they have the same variable values.

We use the hyperplane $h = ****0*0***0***0**0$ to decompose the VIG in order to produce a *Recombination Graph*. We remove all the variables (vertices) that have the same "shared variable assignments" and also remove all edges that are incident on the vertices corresponding to these variables. This produces the recombination graph shown in Fig. 2.

The recombination graph also defines a reduced evaluation function. This new evaluation function is linearly separable, and decomposes into q subfunctions defined over the recombining components. In our example:

$$g(x') = a + g_1(9, 11, 16) + g_2(0, 1, 2, 5) + g_3(3, 7, 8, 12, 13, 15),$$

where $g(x') = f|_h(x')$ and x' are restricted to a subspace of the hyperplane h that contains the parent strings as well as the full dynastic potential. The constant $a = f(x') - \sum_{i=1}^{3} g_i(x')$ depends on the common variables.

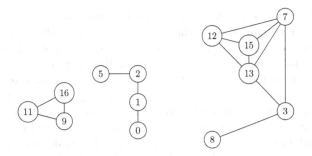

Fig. 2. Recombination Graph for the solutions (parents) $red = 000000000000000000$ and $blue = 111101011101110110$.

The Partition Crossover operator (PX), defined by Tinós et al. [6] is based on this recombination graph. Every recombination graph with q connected components induces a new *separable* function $g(x')$ that is defined as:

$$g(x') = a + \sum_{i=1}^{q} g_i(x'). \tag{2}$$

Partition Crossover selects the parent yielding the best partial solution for each subfunction $g_i(x')$. All of the variables in the same recombining component in the recombination graph must be inherited together from one of the two parents.

Articulation Points Partition Crossover (APX) [7] goes further and finds the *articulation points* of the recombination graphs. They are variables whose removal increases the number of connected components. Variables x_1, x_2 and x_3 are articulation points in our example (see Fig. 2). Then, APX efficiently simulates what happens when the articulation points are removed, one at a time, from the recombination graph by flipping the articulation point in any of the parent solutions before applying PX, and the best solution is returned as offspring. With the appropriate data structures, this can be done in $O(n^2 + m)$, the same complexity of PX.

3   Dynastic Potential Exploration

The proposed Dynastic Potential Crossover Operator (DPX) takes the idea of PX and APX even further. DPX starts from the recombination graph, like the one in Fig. 2, and tries to exhaustively explore all the possible combinations of the parent values in the variables of each connected component to find the optimal recombination regarding the hyperplane h defined by the blue and red parents. This exploration is not done by brute force, but using dynamic programming. Following with our example, in order to compute the best combination for the variables x_9, x_{11} and x_{16}, we need to enumerate the 8 ways of taking each variable from each parent, and this is not better than brute force. However, component x_0, x_1, x_2, x_5, forms a thread. In this case we can store in a table which is the best option for variable x_0 when any of the two possible values for variable x_1 are selected and we can store in the same table what is the value of the sum of subfunctions depending only on x_0 and x_1 (and possibly common variables eliminated in the recombination graph). After this step, we can consider that variable x_0 has been removed from the problem and we can proceed in the same way with the rest of the variables in the order x_1, x_2 and x_5. At the end, only 12 evaluations are necessary, while a brute force would require 16 evaluations.

The idea of variable elimination using dynamic programming dates back to the 1960's and Hammer's basic algorithm [3]. It is well-known that the complexity of this approach is $O(N2^t)$, where t is the treewidth of the graph. Computing the treewidth of a graph is an NP-hard problem [4]. Thus, heuristics should be applied to find an elimination order for the variables. The problem of variable elimination has also been studied in other contexts, like Gaussian Elimination [12] and Bayesian Networks [4]. In fact, we follow the works done for computing the *junction tree* in Bayesian Networks. In order to do this, we first need a *chordal graph* and then compute the *clique tree* (or junction tree), which will fix the order in which the variables are eliminated using Dynamic Programming. Our contribution in this work consists in applying these ideas to the recombination operator. The high level pseudocode of the proposed DPX is presented in Algorithm 1. In the next subsections we will detail each of these steps.

Algorithm 1. Pseudocode of DPX

Input: two parents x and y
Output: one offspring z
  1: Compute the Recombination Graph of x and y as in [6]
  2: Apply Maximum Cardinality Search to the Recombination Graph [12]
  3: Apply the fill-in procedure to make the graph chordal [12]
  4: Apply the Clique Tree construction procedure [13]
  5: Assign subfunctions to cliques in the clique tree
  6: Apply Dynamic Programming to find the offspring (see Algorithm 2)
  7: Build z using the tables filled by Dynamic Programming

3.1 Chordal Graphs

A *chordal graph* is a graph where all the cycles of length 4 or more have a chord (edge joining two nodes not adjacent in the cycle). All the connected components in Fig. 2 are chordal graphs. Tarjan and Yannakis [12] provided algorithms to test if a graph is chordal and add new edges to make it chordal if it is not. Their algorithms run in time $O(n + e)$, where e is the number of edges in the graph. In the worst case the complexity is $O(n^2)$. The first step to check the chordality is to number the nodes using *Maximum Cardinality Search* (MCS). This algorithm numbers each node in descending order, choosing always the unnumbered node with a higher number of numbered neighbors and solving the ties arbitrarily. Figure 3 (left) shows the result of applying MCS to the third connected component of Fig. 2.

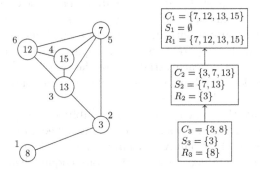

Fig. 3. Maximum Cardinality Search applied to the third connected component of Fig. 2 (left) and clique tree with the sets of separators and residues (right).

If the graph is chordal then MCS will provide a numbering of the nodes such that for each triple of nodes u, v and w, with $(u, v), (u, w) \in E$ and u has a lower number than v, w, it happens that $(v, w) \in E$. If this is not the case, the graph is not chordal. A *fill-in* algorithm tests this condition and adds the required edges to make the graph chordal. This algorithm runs in $O(n + s')$ time, where s' is the number of edges in the final chordal graph. Again, in the worst case, the complexity is $O(n^2)$. These two steps, MCS and fill-in, can be computed to each connected component separately or to the complete recombination graph with the same result. The algorithms are applied in Lines 2 and 3 of Algorithm 1.

3.2   Clique Tree

Dynamic Programming is based on the exhaustive exploration of the cliques[1] in the chordal graph. The maximum size of a clique in the chordal graph is an upper bound of its treewidth, and determines the complexity of applying dynamic programming to find the optimal solution. A clique tree of a chordal graph is a tree where the nodes are cliques and for any variable appearing in two of such cliques, the path among the two cliques in the tree is composed of cliques containing the variable (junction tree property). We can also identify a clique tree with a tree-decomposition of the chordal graph [4]. This clique tree will determine the order in which the variables can be eliminated.

Starting from the chordal graph provided in the previous steps, we apply an algorithm by Galinier et al. [13] to find the clique tree (Line 4 in Algorithm 1).

---

[1] We will use the term *clique* to refer to a maximal complete subgraph, as the cited literature does. However, the term clique is sometimes used to refer to a complete subgraph (not necessarily maximal).

This algorithm runs also in $O(n + e')$ time and finds all the $O(n)$ cliques of the chordal graph. The cliques will be denoted with C_i, where i is an index that increases when a clique is discovered by the algorithm. An edge joining two cliques in the clique tree is labelled with a *separator*, which is the intersection of the variables in both cliques. A clique C_i is parent of a clique C_j if they are joined by an edge and $i < j$. In each clique C_i the *residue*, R_i, is the set of variables that are not in the separator with its parent. In each clique C_i, the residue, R_i, and the separator with the parent, S_i, forms a partition of the variables in C_i. It is not hard to prove that each variable is in the residue of one clique only. In Fig. 3 (right) the residues and separators for all the cliques of the third connected component of Fig. 2 are shown.

After computing the clique tree, all the subfunctions f_l depending on a nonempty set V of differing variables must be assigned to one (and only one) clique C_i containing V (Line 5 in Algorithm 1). They will be evaluated when this clique is processed. There can be more than one clique where the subfunction can be assigned. All of them are valid for a correct evaluation. We denote with F_{C_i} the set of subfunctions assigned to clique C_i.

The optimal offspring is found by iteratively reducing the variables in the residue of the cliques (Line 6 in Algorithms 1 and 2). The clique tree must be traversed in post-order in order to do this. During the clique evaluation, for each combination of variables in the separator S_i (Line 2 in Algorithm 2), all the combinations of variables in the residue R_i are considered (Line 4 in Algorithm 2) and evaluated over the subfunctions assigned to the clique (Lines 6–8) and their child cliques (Lines 9–11). The evaluation in post-order makes it possible to have the value array of the child cliques filled when they are evaluated. The best combination of the variables in R_i for each combination of the variables of S_i is stored in the array variable in Line 14. This array will be used in the reconstruction of the offspring solution (Line 7 in Algorithm 1). In Algorithm 2 we assume that value 0 for a variable means the value in the red parent and a 1 means the value in the blue parent. The term x_V for V a set of variables, will denote a vector with the variables in V.

The operator described is an Optimal Recombination operator: it finds the best offspring from the largest dynastic potential. The time required to evaluate one clique in Algorithm 2 is $O((|F_{C_i}| + |children(C_i)|)2^{|C_i|})$, where $children(C_i)$ is the set of child cliques of C_i. The number of children is bounded by n and the number of subfunctions m is bounded by $O(n^k)$ due to the k-bounded epistasis of f. However, the exponential factor is a threat to the efficiency of the algorithm. In the worst case C_i can contain all the variables and the factor would be 2^n.

3.3   Limiting the Complexity

In order to avoid the exponential runtime, we propose to limit the exploration in Lines 2 and 4. Instead of iterating over all the possible combinations for all the variables in S_i and R_i we fix a bound β on the number of variables that will be exhaustively explored. The remaining variables will jointly take only two values, each one coming from one of the parents. This reduces the exponential part of the

---

Algorithm 2. Optimal Offspring Computation

```
 1: for all cliques C_i of the clique tree in post-order do
 2:     for x_{S_i} ∈ {0,1}^{|S_i|} do
 3:         value[x_{S_i}] = -∞
 4:         for x_{R_i} ∈ {0,1}^{|R_i|} do
 5:             aux = 0
 6:             for f ∈ F_{C_i} do
 7:                 aux = aux + f(x)
 8:             end for
 9:             for children cliques C' of C_i do
10:                 aux = aux + value[x_{C'}];
11:             end for
12:             if aux > value[x_{S_i}] then
13:                 value[x_{S_i}] = aux
14:                 variable[x_{S_i}] = x_{R_i}
15:             end if
16:         end for
17:     end for
18: end for
```

---

complexity of Algorithm 2 to $2^{2\beta}$. Since β is a predefined constant decided by the user of the algorithm, the exponential factor turns into a constant. The operator is not anymore an optimal recombination operator, and this is the reason why we call it *quasi-optimal*. In the cases where $\beta \geq |C_i|$ for all the cliques, the operator will still return the optimal offspring. The next theorem presents the complexity of DPX.

Theorem 1. *Given a function in the form of* (1) *with m subfunctions, the complexity of DPX with a constant bound β for the number of exhaustively explored variables is $O(4^\beta(n+m)+n^2)$.*

Proof. We have seen in Sect. 3.1 that the complexity of Maximum Cardinality Search, the fill-in procedure and the clique tree construction is $O(n^2)$. The assignment of subfunctions to cliques can be done in $O(n+m)$ time, using the variable ordering found by MCS to assign the subfunctions that depends on each visited variable to the only clique where the variable is a residue. The complexity of the dynamic programming computation is:

$$O\left(\sum_i (|F_{C_i}| + |children(C_i)|)2^{|C_i|}\right) = O\left(2^{2\beta}\sum_i (|F_{C_i}| + |children(C_i)|)\right)$$

$$= O\left(4^\beta(m + \sum_i |children(C_i)|)\right)$$

$$= O(4^\beta(m+n)),$$

where we used the fact that the sum of the cardinality of the children for all the cliques is the number of edges in the clique tree, which is the number of cliques minus one, and the number of cliques is $O(n)$. The reconstruction of the offpsring solution requires to read all the `variable` tables until building the solution. The complexity of this procedure is $O(n)$.   □

In many cases, the number of subfunctions m is $O(n)$ or $O(n^2)$. This is true, in particular, when the function has k-bounded epistasis. In these cases, the complexity of DPX reduces to $O(4^\beta n^2)$.

3.4   Theoretical Comparison with (A)PX

It is clear that DPX is no worse than PX, since it considers each connected component in the recombination graph and, in the worst case, it will do the same as PX and will pick the variables from one of the parent solutions. We wonder, however, if this happens with APX. If β is large enough for a given recombination, it cannot be worse than any recombination operator with the property of gene transmission and, in particular, cannot be worse than APX. If β is not that large and the limit in the exploration (SubSect. 3.3) is applied, it could happen that articulation points are not explored as they are in APX. One possible threat to the articulation points exploration in DPX is that they disappear after making the graph chordal. The next result proves that articulation points survive the fill-in procedure and inspires a mechanism to reduce the probability that a solution explored in APX is not explored in DPX.

Theorem 2. *Articulation points of a graph are kept after the fill-in procedure.*

Proof. Proving that all articulation points survive the fill-in procedure is equivalent to proving that all the edges added by the fill-in procedure join vertices of one single bi-connected component. If an edge (v, w) is added joining vertices of two different bi-connected components, then two paths would exist to go from v to w: the original path traversing at least one articulation point a and the new edge. But, in this case, the articulation point a could be removed from the graph. In the other direction, adding edges to a bi-connected component never removes articulation points.

We assume that MCS has been applied to the graph. We denote with $\gamma(v)$ the number assigned by MCS to node v. Let us prove the claim by contradiction. Imagine that edge (v, w) is added in the fill-in procedure, where v and w are in different bi-connected components. The definition of fill-in (see [12]) implies that there is a path among v and w where all the intermediate nodes have a γ value lower than v and w. In particular, since v and w are in different bi-connected components, all the paths between them include the same set of articulation points and for all of them the value of γ is lower than $\min(\gamma(v), \gamma(w))$. MCS numbers the nodes in a connected component in decreasing order and in such a way that all the numbered nodes are connected. Thus, in all the bi-connected components the first node numbered by MCS is an articulation point, with the only exception of the bi-connected component where the numbering starts. This implies that in one of the bi-connected components, say the one of v, there is an articulation point a_v with $\gamma(a_v) > \gamma(v)$ that was the first numbered in that bi-connected component. Regarding the bi-connected component of w, if it is the one where the numbering started, then there must be an alternative path from w to v though a_v. But this means that w and v belongs to the same bi-connected component, what is a contradiction. If the bi-connected component of w is not where the numbering started, there must be an articulation point a_w with $\gamma(a_w) > \gamma(w)$ where the numbering started in that bi-connected component. Once again, there must be an alternative path between v and w through a_w and a_v, contradicting the fact that v and w are in different bi-connected components. Then, the fill-in procedure will not add edge (v, w).   □

The previous theorem implies that articulation points of the original recombination graph are also articulation points of the chordal graph. Articulation points of a chordal graph are minimal separators of cardinality 1 (see [13]) and they will appear as separators S_i in some cliques C_i. They are, thus, identified during the clique tree construction. In each clique C_i when β variables are chosen to be exhaustively explored (Lines 2 and 4 of Algorithm 2) we choose the articulation points first. This way, articulation points can be exhaustively explored with higher probability. The only thing that can prevent articulation points from being explored is that many of them appear in one single clique. This situation is illustrated in Fig. 4. For $\beta \leq 1$, the clique of articulation points is evaluated only in the two parent solutions, and the same happens with the other cliques, giving a total of 16 explored combinations. However, APX would explore 20 combinations in this situation (see Eq. (6) in [7]).

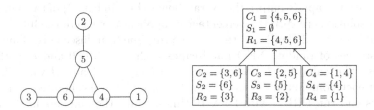

Fig. 4. Pathological component (left) in a recombination graph where DPX with $\beta \leq 1$ explores less solutions (16) than APX (20) and its clique tree (right).

4   Experiments

In order to experimentally analyze the performance of DPX, we included it in the Deterministic Recombination and Iterated Local Search (DRILS) algorithm [7]. We think this allows us to explore the performance of the operator in a real scenario, rather than generating random solutions and providing them to the operator. DRILS [8] uses a first improving move hill climber to reach a local optimum. Then, it perturbs the solution by randomly flipping αN bits, where α is the so-called *perturbation factor*. It then applies local search to the new solution to reach another local optimum and applies crossover to the last two local optima, generating a new solution that is improved further with the hill climber. This process is repeated until a time limit is reached.

In our case, the recombination operator is DPX, but we also present results with PX and APX in SubSect. 4.2 to compare the operators. In all the runs we set a time limit of 60 s (1 min). Since the algorithms are stochastic, we performed 10 independent runs for each instance and algorithm. We tested DRILS with DPX in NP-hard problems: random NKQ Landscapes with $K \geq 2$, which allows us to parameterize the density of edges in the VIG by changing K; and MAX-SAT instances of the MAX-SAT Evaluation 2017. Random NKQ ('Quantized' NK) landscapes [14] can be seen as Mk landscapes with one subfunction per variable ($m = n$). Each subfunction f_l depends on variable x_l and other $K = k-1$ random variables, and the codomain of each subfunction is the set $\{0, 1, \ldots, Q-1\}$, where Q is a positive integer. The values of the subfunctions are randomly generated. Random NKQ landscapes are NP-hard when $K = k - 1 \geq 2$. The computer used for the experiments is a multicore machine with four Intel Xeon CPU (E5-2670 v3) at 2.3 GHz, a total of 48 cores, 64 GB of memory and Ubuntu 16.04 LTS. The source code of all the algorithms can be found at https://github.com/jfrchicanog/EfficientHillClimbers.

4.1 DPX Statistics

In a first experiment, we compute statistics about DPX. In particular, in Tables 1 and 2 we count the average number of connected components identified in the recombination graph (Comp.), the average logarithm (in base 2) of the number of explored solutions (Exp.), the percentage of applications where the full dynastic potential is explored (Full) and the average runtime in milliseconds (Time). We used instances of random NKQ Landscapes with $n = 10\,000$ and $n = 100\,000$ variables. The value for K varies from 2 to 5, $Q = 64$ and β varies from 2 to 5. For each combination of the parameters n and K we generated 10 random instances and run DRILS with DPX 10 times. Thus, the numbers in the tables are averages over 100 runs (the percentage of full explorations counts all the applications of crossover in the 100 runs). The perturbation factor (α) in DRILS was set to $\alpha = 0.05$ in the cases $K = 2, 3$ and $\alpha = 0.01$ in the cases $K = 4, 5$. These values were taken from the recommendations in [8].

We observe in the tables that the percentage of applications of DPX where the full dynastic potential is explored is high, almost always around or above 90%, except in the case of $N = 100\,000$ and $K = 3$. This percentage should increase with β and it normally does, being the exceptions not significant. But the fact that the value is high for low values of β (2 or 3) is an indication that the cliques found in the recombination graph are small, with size 2 or 3 in most of the cases. One can imagine that this corresponds to threads of variables with some triangles sometimes. This corresponds with the plots presented by Chen et al. in [9]. Due to this high percentage of success we can trust that the logarithm

Table 1. DPX Statistics for $n = 10\,000$ variables.

	Comp.	Exp.	Full (%)	Time (ms)		Comp.	Exp.	Full (%)	Time (ms)
$K = 2$					$K = 4$				
$\beta=2$	64	212	99.9	6.8	$\beta=2$	13	44	99.3	6.2
$\beta=3$	64	210	100.0	3.9	$\beta=3$	12	42	100.0	3.5
$\beta=4$	64	210	100.0	3.9	$\beta=4$	12	42	100.0	3.4
$\beta=5$	64	210	100.0	3.9	$\beta=5$	12	42	100.0	3.6
$K = 3$					$K = 5$				
$\beta=2$	50	232	94.7	8.4	$\beta=2$	11	42	99.0	7.1
$\beta=3$	50	225	99.8	4.7	$\beta=3$	11	39	99.9	3.8
$\beta=4$	50	226	99.9	4.7	$\beta=4$	11	39	99.9	3.8
$\beta=5$	50	226	99.9	4.8	$\beta=5$	11	40	100.0	4.0

Table 2. DPX Statistics for $n = 100\,000$ variables.

	Comp.	Exp.	Full (%)	Time (ms)		Comp.	Exp.	Full (%)	Time (ms)
$K = 2$					$K = 4$				
$\beta=2$	668	2 249	99.9	75.5	$\beta=2$	140	567	91.2	69.2
$\beta=3$	668	2 249	100.0	74.1	$\beta=3$	140	566	99.1	68.8
$\beta=4$	668	2 248	100.0	72.5	$\beta=4$	141	571	99.3	71.0
$\beta=5$	670	2 261	100.0	81.6	$\beta=5$	142	587	99.2	82.2
$K = 3$					$K = 5$				
$\beta=2$	505	2 693	63.2	110.6	$\beta=2$	121	570	89.6	75.9
$\beta=3$	505	2 691	94.0	109.6	$\beta=3$	121	570	89.6	75.9
$\beta=4$	506	2 702	94.7	113.3	$\beta=4$	122	575	96.9	77.2
$\beta=5$	505	2 726	94.8	126.1	$\beta=5$	123	596	96.8	91.9

of the number of explored solutions (column "Exp." in the tables) is a good measure of the number of differing variables in the parent solutions. If we divide this number by the number of components we find a value between 3 and 4. This must be the average number of variables in each connected component.

Both the number of components and the differing variables are approximately multiplied by 10 when we compare the 10 K variable instances with the 100 K variable instances. We observe, however, that these values are similar for $K = 2, 3$ and are divided by 4 or 5 when $K = 4, 5$. The reason is the perturbation factor α, which is also divided by 5 in these instances.

The runtime is in the order of a few milliseconds for 10 K variables and 70 to 100 ms for 100 K variables. This runtime should increase with β but we observe some exceptions for low β. The reason has to do with the procedures in DPX used to identify the group of variables that will be exhaustively explored and the one for which only the parent solutions will be evaluated. From the results of the tables, we conclude that a value for β of 3 or 4 is the best one for these instances. Two other parameters affecting the runtime are the perturbation factor α, because it will determine the number of differing variables (the higher the value the higher the runtime), and K, since it will add edges to the recombination graph. This is why we observe that runtime increases from $K = 2$ to $K = 3$ and from $K = 4$ to $K = 5$. Anyway, this runtime is small compared to the number of solutions that are explored. If we take the results for $N =$100K and $K = 3$ as an example, DPX is exploring 2^{2693} solutions in 110 ms. This is equivalent to exploring around 10^{800} solutions per nanosecond (ns) if a black box approach is used.

4.2   Comparison with PX and APX for NKQ Landscapes

In this section we compare DPX with PX and APX. Table 3 shows a comparison regarding three aspects: exploration capacity, runtime and performance inside DRILS. The first two aspects depend only on the crossover operators and the third one depends also on the algorithm (DRILS). For the exploration capacity we show the logarithm in base 2 of the number of explored solutions by each operator. We observe how DPX has the largest exploration capacity, around the square of the one of APX (the logarithm is around double) and between the fourth and fifth power compared to the one of PX. In terms of runtime, DPX requires more time than PX and APX, as expected, and this time is between 20% and 70% higher than PX and APX. Finally, we compare the performance of DRILS using each of the crossover operators. For each instance (ten per value of K) we compare the medians of the algorithms after 1 min of computation and we apply the Mann-Whitney test (with significant level 0.05) to check if the differences are statistically significant. The numbers followed by a black triangle ($\blacktriangle$), white triangle ($\triangledown$) and equal sign ($=$) are the numbers of instances in which DRILS with DPX is statistically better, worse or similar to DRILS with the operator of that column (PX or APX). The performance comparison suggests that DPX is improving the search of DRILS only for $K = 3$. In the other cases the other two operators (specially APX) are better. A complete explanation of this observation requires further research, but we can guess that DPX can be too greedy, providing a solution which is a (near) local optimum difficult to escape from. It also requires more time to run and this time is used in the other versions of DRILS to escape from the local optima. Both ideas can be checked with a Local Optimal Network (LON) analysis, which we defer to future work.

4.3   Comparison with PX and APX for MAX-SAT

In Table 4 we compare PX, APX and DPX using MAX-SAT instances from the MAX-SAT Evaluation 2017[2]. We used the same instances as in [7][3] to allow an easy comparison. They are 160 unweighted and 132 weighted instances.

---

[2] http://mse17.cs.helsinki.fi/benchmarks.html.
[3] The list of instances is at https://github.com/jfrchicanog/EfficientHillClimbers.

Table 3. Comparison of PX, APX and DPX for $N = 100\,000$ variables. The value for α depends on K as described in the text and in DPX we used $\beta = 4$.

K	Exploration			DRILS performance		Runtime (ms)		
	PX	APX	DPX	PX	APX	PX	APX	DPX
2	662	1311	2248	1▲ 0▽ 9 =	0▲ 8▽ 2 =	46	55	73
3	503	1105	2702	10▲ 0▽ 0 =	2▲ 0▽ 8 =	73	67	113
4	138	286	571	0▲ 4▽ 6 =	0▲ 9▽ 1 =	52	55	71
5	119	254	575	0▲ 9▽ 1 =	0▲ 10▽ 0 =	52	63	77

Table 4. Comparison of PX, APX and DPX for MAX-SAT instances (weighted and unweighted). In all the cases $\alpha = 0.3$ in DRILS and $\beta = 4$ in DPX.

Instances	DRILS performance		Runtime (μs)		
	PX	APX	PX	APX	DPX
Unweighted	126▲ 2▽ 32 =	96▲ 19▽ 45 =	1060	849	1907
Weighted	102▲ 14▽ 16 =	90▲ 17▽ 25 =	1713	2365	5171

We observe how DPX required on average twice the time required by PX and APX in each run, in the order of 2 to 5 ms. However, the performance of DRILS using DPX is significantly better in most of the instances than the performance when PX or APX is used. In particular, DPX is statistically better than APX and PX in 96 and 126 unweighted instances, respectively. The difference in the weighted instances is not so high, but still large enough to be promising. Interestingly, weighted MAX-SAT instances must have a fitness landscape similar to NKQ Landscapes while unweighted instances have a different fitness landscape, with many plateaus difficult to escape from. DRILS with DPX seems to work better than DRILS with PX and APX in such a plateau-based landscape. We defer to future work the detailed analysis of this performance and the big difference with NKQ Landscapes.

5 Conclusions

In this paper we propose a new gray box crossover operator, DPX, with the ability to obtain the best offspring out of the full dynastic potential if the density of interactions among the variables is low. We have provided theoretical results proving that DPX is no worse than Partition Crossover (PX) and usually no worse than Articulation Points Partition Crossover (APX). We also compared these three operators inside the DRILS algorithm in NKQ Landscapes and MAX-SAT, certifying its exploration ability.

An interesting future line of research is to analyze the operator using Local Optima Networks and the shape of the connected components of the recombination graph to understand the reasons for the observed different performance in NKQ Landscapes and MAX-SAT. It would also be interesting to check the performance of the operator in other algorithms and to develop a competitive MAX-SAT solver based on it.

References

1. Radcliffe, N.J.: The algebra of genetic algorithms. Ann. Math. Artif. Intell. **10**(4), 339–384 (1994)
2. Eremeev, A.V., Kovalenko, J.V.: Optimal recombination in genetic algorithms. CoRR abs/1307.5519 (2013). http://arxiv.org/abs/1307.5519
3. Hammer, P.L., Rosenberg, I., Rudeanu, S.: On the determination of the minima of pseudo-boolean functions. Stud. Cerc. Mat. **14**, 359–364 (1963)
4. Bodlaender, H.L.: Discovering treewidth. In: Vojtáš, P., Bieliková, M., Charron-Bost, B., Sýkora, O. (eds.) SOFSEM 2005. LNCS, vol. 3381, pp. 1–16. Springer, Heidelberg (2005). https://doi.org/10.1007/978-3-540-30577-4_1
5. Whitley, D., Chicano, F., Goldman, B.W.: Gray box optimization for mk landscapes (nk landscapes and max-ksat). Evol. Comput. **24**, 491–519 (2016)
6. Tinós, R., Whitley, D., Chicano, F.: Partition crossover for pseudo-boolean optimization. In: Proceedings of the 2015 ACM Conference on Foundations of Genetic Algorithms XIII FOGA 2015, pp. 137–149. ACM, New York (2015)
7. Chicano, F., Ochoa, G., Whitley, D., Tinós, R.: Enhancing partition crossover with articulation points analysis. In: Proceedings of GECCO. GECCO 2018, pp. 269–276. ACM, New York (2018)
8. Chicano, F., Whitley, D., Ochoa, G., Tinós, R.: Optimizing one million variable NK landscapes by hybridizing deterministic recombination and local search. In: Genetic and Evolutionary Computation Conference GECCO 2017, pp. 753–760 (2017)
9. Chen, W., Whitley, D., Tinós, R., Chicano, F.: Tunneling between plateaus: improving on a state-of-the-art maxsat solver using partition crossover. In: Proceedings of GECCO. GECCO 2018, pp. 921–928. ACM, New York (2018)
10. Tinós, R., Zhao, L., Chicano, F., Whitley, D.: Nk hybrid genetic algorithm for clustering. IEEE Trans. Evol. Comput. **22**(5), 748–761 (2018)
11. Terras, A.: Fourier Analysis on Finite Groups and Applications. Cambridge University Press, Cambridge (1999)
12. Tarjan, R.E., Yannakakis, M.: Simple linear-time algorithms to test chordality of graphs, test acyclicity of hypergraphs, and selectively reduce acyclic hypergraphs. SIAM J. Comput. **13**(3), 566–579 (1984)
13. Galinier, P., Habib, M., Paul, C.: Chordal graphs and their clique graphs. In: Nagl, M. (ed.) WG 1995. LNCS, vol. 1017, pp. 358–371. Springer, Heidelberg (1995). https://doi.org/10.1007/3-540-60618-1_88
14. Newman, M.E.J., Engelhardt, R.: Effect of neutral selection on the evolution of molecular species. Proc. R. Soc. London B **256**, 1333–1338 (1998)

Insights into the Feature Selection Problem Using Local Optima Networks

Werner Mostert[1](✉), Katherine M. Malan[2](✉), Gabriela Ochoa[3](✉),
and Andries P. Engelbrecht[1](✉)

[1] Stellenbosch University, Stellenbosch Central, Stellenbosch, South Africa
werner.mostert1@gmail.com, engel@sun.ac.za
[2] University of South Africa, Muckleneuk, Pretoria 0002, South Africa
malankm@unisa.ac.za
[3] University of Stirling, Stirling FK9 4LA, UK
gabriela.ochoa@cs.stir.ac.uk

Abstract. The binary feature selection problem is investigated in this paper. Feature selection fitness landscape analysis is done, which allows for a better understanding of the behaviour of feature selection algorithms. Local optima networks are employed as a tool to visualise and characterise the fitness landscapes of the feature selection problem in the context of classification. An analysis of the fitness landscape global structure is provided, based on seven real-world datasets with up to 17 features. Formation of neutral global optima plateaus are shown to indicate the existence of irrelevant features in the datasets. Removal of irrelevant features resulted in a reduction of neutrality and the ratio of local optima to the size of the search space, resulting in improved performance of genetic algorithm search in finding the global optimum.

Keywords: Local optima networks · Feature selection ·
Fitness landscape analysis

1 Introduction

To further the development of a generalised theoretical framework for feature selection, and to better understand the feature selection problem, this paper analyses fitness landscapes of the feature selection problem.

The binary feature selection problem has the goal of finding the subset of all features that are the most relevant for a classification task. A full enumeration of candidate solutions (i.e subsets of features) is performed to construct a complete fitness landscape for a number of real-world classification problems. The size of the solution space grows exponentially as the number of features increases, making a full enumeration computationally infeasible for a large number of features. Therefore, only datasets with a small number of features (less than 18) are considered in this paper.

© Springer Nature Switzerland AG 2019
A. Liefooghe and L. Paquete (Eds.): EvoCOP 2019, LNCS 11452, pp. 147–162, 2019.
https://doi.org/10.1007/978-3-030-16711-0_10

A number of different approaches have been proposed for solving the feature selection problem [4,12–14,23,24]. Chandrashekar et al. [4] showed that the performance of different feature selection techniques are often problem dependent and that the methods show vast disparity in their success ratios. This paper uses two wrapper methods and one filter method to analyse the behaviour of feature selection algorithms with respect to problem landscape characteristics. Fitness landscapes can be a valuable tool to understand problems and to analyse search algorithm behaviour [20], since fitness landscapes possess structural attributes that influence the performance of search algorithms [16].

Local optima networks [19] have previously been used to characterise the global structure of fitness landscapes for benchmark combinatorial problems. The quadratic assignment problem [6], NK-landscapes [22] and the number partitioning problem [18] are examples of combinatorial problem case studies using local optima networks as an analytical tool. For small instances of the feature selection problem (in this study with less than 18 features), a full enumeration of all candidate solutions is computationally practical. This allows for the construction of complete local optima networks and a full analysis of the global structure of fitness landscapes for instances of classification problems.

New insights to the nature of the feature selection problem are obtained using local optima networks. The local optima networks for the real world classification problems reveal interesting fitness landscape characteristics of the feature selection problem, before and after feature removal. Removal of irrelevant features shows a reduction in neutrality in the fitness landscape, a reduction in the ratio of local optima to the size of the search space, and a reduction in problem difficulty for genetic algorithm search.

The following section gives an overview of the feature selection problem and local optima networks, while Sect. 3 describes the experimental process and algorithm details. Finally, the results obtained are discussed in Sect. 4 and the paper is concluded in Sect. 5.

2   Background and Related Work

This section discusses the general feature selection problem and gives an overview of local optima networks.

2.1   The Feature Selection Problem

The feature selection problem is concerned with finding a set of the most relevant features from the set of all available features for a classification task. A solution to the feature selection problem can be represented as a binary string of length n, where n is the number of features in the dataset and each bit indicates whether the feature is selected or not.

Feature selection is applied as a pre-processing technique in order to reduce the dimensionality of a problem by removing redundant and irrelevant features. The issue of feature irrelevance can be misleading since two mutually exclusive features could be useless, but the union of these features could be information rich with respect to the dependant variable [10]. The utilisation of a subset of relevant features as opposed to the set of all features has been shown to increase classifier performance, reduce computational complexity, and lead to a better understanding of the data for machine learning [4].

Feature selection algorithms can be categorised into three distinct categories, namely filter, wrapper and embedded methods. Filter methods [4] establish how important features are based on information with respect to the dependent variable, using measures such as correlation or mutual information. Wrapper methods [4] use subsets of features, for which a measure of the model accuracy is obtained per subset of features. Heuristic search is used by wrapper methods to determine the set of most relevant features with the model accuracy as objective function. Embedded methods [4] search for the most relevant features by taking advantage of the built in learning process, such as with decision trees [10].

Choosing the most suitable feature selection algorithm for a given dataset is an unsolved problem, since there is a lack of an underlying theoretical framework and a limited understanding of the nature of the feature selection problem [10]. Fitness landscapes can be used to better understand the nature of the search spaces of optimisation problems [16] and in this paper, local optima networks are employed to characterise and visualise the fitness landscapes for the feature selection problem.

2.2   Local Optima Networks

A fitness landscape is formulated as a triplet (S, N, f) [21], where S is the set of all candidate solutions, N is an operator that defines a neighbourhood structure in the solution space, and $f : S \longrightarrow \mathbb{R}$ is a fitness (objective) function that assigns to all $s \in S$ a solution quality. Binary solution spaces are have a size of $|S| = 2^N$, where N represents the dimensionality of the problem.

Local optima networks [19] are inspired by the work of Doye [7] in modelling physical energy landscapes of atomic clusters as complex networks, serving as a powerful tool to analyse the global structure of a fitness landscape. For combinatorial spaces, the idea is adapted from connected energy minima to weighted graphs of local optima in the fitness landscape [19]. Definitions of the components of the local optima network model are given below, assuming a maximisation problem.

Local Optimum. A local optimum $s^* \in S^*$, where $S^* \subset S$ is the set of locally optimum solutions, is defined as a solution where $\forall s \in N(s^*), f(s) < f(s^*)$.

Local Optima Network. A local optima network, $G = (S^*, E)$, represents the weighted graph of all local optima solutions S^*, where an escape edge $e_{ij} \in E$ exists between nodes S_i^* and S_j^*.

Escape Edge. An escape edge [22] represents the probability of moving into a neighbouring basin of attraction from a local optimum, after a defined controlled perturbation followed by a hill-climbing local search to find the connecting local optima.

Basin of Attraction. A basin of attraction of a local optimum s_i^* is the set $b_i = \{s \in S \mid LocalSearch(s) = s_i^*\}$. The number of solutions, $|b_i|$, in the basin of attraction represents the size of the basin.

Monotonic Local Optima Network (M-LON). The monotonic local optima network, $M\text{-}LON = (S^*, E)$, is the graph where the nodes $s_i^* \in S^*$ are the local optimum plateaus, and there is an edge $e_{ij} \in E$, with weight w_{ij}, between two nodes s_i^* and s_j^* if $w_{ij} > 0$.

M-LON Plateau. The set of connected nodes in the *M-LON* with equal fitness.

Compressed LON Nodes. The set of *M-LON* plateaus, C^*.

Compressed Monotonic Local Optima Network (CM-LON). The local optima network, $CM\text{-}LON = (C^*, E)$, is the graph where the nodes $c_i \in C^*$ are *M-LON* plateaus. Weighted edges in the *CM-LON* are aggregated for the edges of nodes in the *M-LON* plateau.

3   Experimental Setting

This section describes the classification datasets used in this study. The construction of the fitness landscape using the full enumeration of feature subsets is then described, followed by the different feature selection algorithms and the approach used to construct and visualise the local optima networks.

3.1   Datasets

The University of California, Irvine (UCI), Machine Learning Repository [15] contains a wide range of datasets that can be used for various machine learning objectives. Seven of the UCI repository classification datasets, containing a variety of nominal and numerical features, were used in this study and are summarised in Table 1.

Table 1. Datasets

Dataset	Nominal	Numerical	# Classes	# Data elements	# Features
breast-cancer	Yes	No	2	286	9
zoo	Yes	Yes	7	101	17
page-blocks	No	Yes	5	5473	10
vowel	Yes	Yes	11	990	13
breast-w	No	Yes	2	699	9
heart-statlog	No	Yes	2	270	13
diabetes	No	Yes	2	768	8

The number of features for the datasets considered were kept small in order to be able to compute a full enumeration of the search space, taking into account that the search space expands exponentially with every feature that is considered.

3.2 Fitness Function

A measure of classification accuracy based on a test dataset is used as the fitness value for a solution, s, where s is a subset of features. A bit string representation is used for solutions to indicate inclusion or exclusion of a specific feature. The test dataset is obtained by doing a 50/50 split of the original dataset, where the training dataset is used to perform the feature selection.

The classic k-nearest-neighbour [1] using Euclidean distance, a simple non-stochastic classifier, is used as implemented in the Weka Machine Learning software development kit [11]. Stochasticity involved with calculating fitness is avoided since it would introduce noise into the fitness landscape. The best value for k is problem dependant. For the purpose of this paper, it is not necessary to use optimal values for k since the focus of this paper is to understand the effect of the inclusion and exclusion of features on the fitness landscape. Therefore, k is arbitrarily chosen as $k = 3$.

The measure of classification accuracy is defined as Cohen's Kappa-statistic, a measure non-biased by class imbalance. Cohen's Kappa is defined as [3]:

$$\kappa = \frac{P_0 - P_c}{1 - P_c} \tag{1}$$

where P_c is the probability of agreement by chance and P_0 is the total agreement (i.e. the number of correctly classified instances). The Kappa statistic allows for the level of agreement with respect to each class label to be measured. The Kappa statistic is a robust measure of accuracy, normalised to the range $[-1, 1]$, where total disagreement is represented as $\kappa = -1$, completely random classification as $\kappa = 0$, and total agreement as $\kappa = 1$. Using non-biased measures of classifier accuracy is important since a raw percentage of correct classification may be statistically biased due to a skewed distribution of classes.

A full enumeration of candidate solutions is done. For datasets with a large number of features, it becomes computationally infeasible to do a full enumeration (an NP-hard problem [2]). A full enumeration for small binary spaces allows the construction of a complete local optima network to analyse and visualise the feature selection problem.

3.3 Feature Selection Algorithms

This section describes the filter method and the two wrapper methods that are used to conduct feature selection in this study. The three feature selection methods below are chosen since each method addresses the problem in a very different way.

Filter Method. The information gain [5] feature evaluation measurement is selected for use by the ranker filter method, as implemented in Weka [11]. Using the training dataset, the features are ranked based on relevance with respect to the class, from high to low. Given a list of ranked features, each linear combination of features from highest relevance to lowest relevance is considered and the fitness of the solution is evaluated.

For example, for the set of all features sorted by decreasing information gain, $F = \{1, 5, 2, 4, 3\}$, the following five feature sets are considered in order: $\{1\}$, $\{1, 5\}$, $\{1, 5, 2\}$, $\{1, 5, 2, 4\}$, and $\{1, 5, 2, 4, 3\}$. The fitness of each subset of features is computed by applying k-NN on the test dataset with the subset of features and using the Kappa statistic as an accuracy measure. The feature set with the highest fitness value is selected as the output of the filter method. The data instances used to determine information gain are restricted to the training set only. This is done to fairly evaluate the performance against the wrapper methods, which use the same set for training.

Sequential Forward Selection Wrapper Method. The sequential forward selection (SFS) algorithm is used as implemented by GreedyStepwise in Weka [11]. The SFS algorithm starts with an initial empty set of features and sequentially adds a feature that results in better fitness up until there are no features that can be added that will result in better fitness than the current solution.

Genetic Algorithm Wrapper Method. The genetic algorithm wrapper method uses the classic genetic algorithm as described by Goldberg [9]. The following parameters were set for the genetic algorithm:

– Population size : 20
– Number of generations : 20
– Crossover probability : 0.6
– Mutation probability : 0.033

The Weka [11] default parameters were used for all datasets. Since the genetic algorithm has a stochastic element, fitness is reported as the mean over 30 independent runs of the algorithm.

3.4 Local Optima Network Generation and Visualisation

The open source Java library by Fieldsend [8] was used to generate the local optima network (LON) graph. Since a full enumeration of the candidate solutions in the search space is done, an exhaustive LON is generated which indicates all of the local and global optima that exist for each problem. In the LON graph, therefore, each node is either a local or a global optimum.

An edge between two nodes in the LON indicates that the two basins containing the optima are regarded as neighbours in the search space. Two common concepts used to define neighbourhood between basins in a LON are basin transitions [17] and escape edges [22]. In this study, the escape edges definition of neighbourhood is used since the basin transition definition, for a full enumeration, produces vastly dense networks. Using escape edges, two local optima are defined as neighbours in the LON if it is possible for any candidate solution in the basin of one local optimum to reach a candidate solution in the basin of the other optimum after a controlled perturbation. In this study, the 'controlled perturbation' for defining neighbourhood between basins is set to 2 bit flips. A larger value for the number of bit flips results in a less connected LON, whereas a single bit flip results in a more densely connected LON.

The generated local optima network is visualised, with the size of the nodes proportional to the basin sizes for the local optima. The edge weights, represented as the width of the edges in the visualisation, between nodes indicates the probability of moving to the connected node.

The nodes and edges are coloured as described in Table 2.

Table 2. Local optima network node & edge colours

Node colour	Node description	Edge colour	Edge description
Red	Global optima	Green	Deteriorating fitness
Blue	Best local optima in a non-global funnel	Blue	Neutral fitness
Gray	All other local optima	Gray	Improving fitness

Compressed monotonic local optima networks [18], as defined in Sect. 2.2, is an adaptation of the local optima networks for connected local optima with equal fitness. Compressed monotonic local optima networks are used for visualisation of neutrality for the feature selection fitness landscapes. The compressed local optima networks show the size of the local optima plateaus as a box, where the length of box is proportional to the size of the plateau.

4   Results

The feature selection algorithm performance for each respective dataset is summarised in Table 3. The performance ratio, P_{ratio}, is calculated as,

$$P_{ratio} = 1 - \frac{f(s_{global} - f(s_{best}))}{s_{global}}$$

where f is the fitness function using Cohen's Kapa statistic for k-KNN, s_{global} is the solution with the best fitness of all candidate solutions (obtained by a full enumeration) and s_{best} is the solution that is returned by the feature selection algorithm. A performance ratio of 1.0 means that the algorithm found a global optimum. For the genetic algorithm (GA) wrapper method, the mean of the fitness values for each run of the algorithm is used to calculate the performance ratio. The GA success ratio is presented as the number of times a single run of the genetic algorithm found a global optimum as a proportion of the number of independent runs (30) of the algorithm. The values in bold show the best performance per feature selection algorithm.

The sequential forwards selection wrapper (SFS) method found a global optimum for two of the datasets. The GA wrapper method found a global optimum for at least one of the runs for all datasets, evident since the success ratio is always greater than 0. The filter method comes close to finding the global optimum, as can be seen in the breast-w dataset with a performance ratio of 0.9920, but failed to find a global optimum for any of the datasets.

Table 3. Feature selection algorithm performance

Dataset	$P_{ratio}(Filter)$	$P_{ratio}(SFS)$	$P_{ratio}(GA)$	GA success ratio
breast-cancer	0.8315	0.9827	0.9758	0.3666
zoo	0.9711	0.8281	0.9872	0.1333
page-blocks	0.9409	**1.0**	0.9896	0.0666
vowel	0.8228	**1.0**	0.9826	0.7
breast-w	**0.9920**	0.9845	**0.9992**	**0.8666**
heart-statlog	0.9443	0.9443	0.9626	0.2
diabetes	0.9284	0.8899	0.9682	0.2333

The GA wrapper method comes close to finding a global optimum in terms of fitness, but shows a largely disparate success ratio with 0.1333 for the vowel dataset and 0.8666 for the breast-w dataset.

Table 4 summarises metrics calculated for the local optima networks for each of the datasets. The metrics give an analytical indication to the structure of the local optima networks. The values in bold represent the highest value for each metric for the different data sets that are analysed.

Table 4. Local optima network metrics

Dataset	N_optima	N_global	Bsize_global	Str_global	N_coptima	Prop_neutral
breast-cancer	30	1	0.0605	0.2252	24	0.2
zoo	**4938**	2	0.0012	0.0094	**1342**	**0.7282**
page-blocks	11	1	0.0958	0.152	11	0
vowel	42	**4**	0.1472	**0.4988**	15	0.6428
breast-w	40	3	0.1192	0.2423	14	0.65
heart-statlog	135	1	0.0135	0.062	109	0.1925
diabetes	8	1	**0.1607**	0.354	8	0

Table 4 contains the following metrics, with definitions as follows:

- *N_optima*: Number of nodes (total number of optima, including global and local)
- *N_global*: Number of different global optima
- *Bsize_global*: Normalised size of the global optimum. Notice that there can be more than one global optimum
- *Str_global*: Incoming strength of global optimum normalised by the total incoming strength
- *N_coptima*: Number of compressed local optima
- *Prop_neutral*: Proportion of local optima that have a neutral connection $\left(\frac{N\_optima - N\_coptima}{N\_optima} \right)$

Three of the datasets exhibited multiple global optima, where the remainder of the datasets exhibited a single global optimum. The dataset that contained the most global optima is the vowel dataset with four global optima. The three datasets that exhibited more than one global optimum also show the largest proportion of neutrality between nodes in the local optima network. The zoo local optima network shows a 0.7282 local optima neutrality proportion, which indicates that 72.82% of all optima in the local optima network have at least one local optimum to which it is connected via a neutral edge (i.e. equal fitness between connected nodes).

The number of local optima are evidently correlated with the size of the search space, but does not necessarily tend to the same exponential scale since the ratio of global optima with regards to the number of candidate solutions differ.

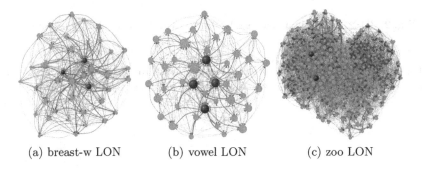

(a) breast-w LON   (b) vowel LON   (c) zoo LON

Fig. 1. Local optima networks

The constructed local optima networks are shown in Fig. 1. Only the datasets with multiple global optima are presented. Due to the size of the zoo local optima network, it has been pruned to 258 out of the 4938 with a fitness value threshold for visualisation purposes.

It is observed from the local optima network visualisations in Fig. 2 that there are multiple global optima that are closely co-located. The nodes that are global optima form a fully connected sub-graph with neutral edges between them. Figure 2 shows a three dimensional visualisation of the local optima network that clearly illustrates the neutral plateau formed by the global optima for the breast-w dataset. The global optima neutral plateau can be seen in Fig. 2 by the three red nodes (global optima) that are fully connected with blue (neutral) edges.

To investigate the presence of other (local) optima plateaus in the LON, a compressed monotonic LON (*CM-LON*) representation is used. This representation shows interconnected local optima via neutral edges as a box in the local optima network, where the length of the box is representative of the number of local optima that are connected. Nodes are still represented as circles where there are no local optima connected via a neutral edge. The *CM-LON* for each

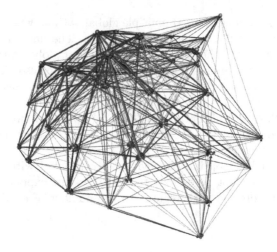

Fig. 2. 3D Local Optima Network for breast-w dataset. (Color figure online)

of the three datasets with multiple global optima are presented in Fig. 3. The colouring scheme remains the same as for the full local optima networks.

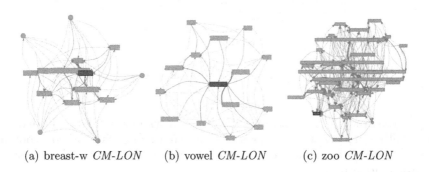

(a) breast-w *CM-LON*   (b) vowel *CM-LON*   (c) zoo *CM-LON*

Fig. 3. Compressed monotonic local optima networks.

Note that the individual nodes of the global optima are now collapsed into a single node, confirming the observation that the global optima are connected via neutral edges. The breast-w dataset exhibits a proportionally large local optima plateau and several smaller local optima plateaus, where the zoo dataset exhibits multiple proportionally larger local optima plateaus. Neutrality in a landscape represents a lack of information for search and metaheuristic search algorithms, such as genetic algorithms. Genetic algorithms require strategies for coping with high levels of neutrality. The vowel *CM-LON* shows that there is no local optimum in the network that does not have at least one other connected local optimum with equal fitness.

The formation of global optima plateaus shows that there are features in the global optima solutions that have no effect on fitness. Therefore, the symmetric intersection of the feature subsets will result in the set of irrelevant features. Table 5 shows the bit string representation of 10 local optima with the highest fitness represented as s_i^* where $i = 1, ..., 10$ is decreasing in fitness. The bits in bold represent the irrelevant features.

Table 5. Feature subsets for top 10 highest fitness local optima.

zoo		breast-w		vowel	
Solution	Features	Solution	Features	Solution	Features
s_1^*	00111101100011001	s_1^*	110**0**011**10**	s_1^*	**00**1111110**0100**
s_2^*	00111101100011011	s_2^*	110**1**010**11**	s_2^*	**0**1111111**00100**
s_3^*	01111101010011001	s_3^*	110**1**011**11**	s_3^*	1011111100100
s_4^*	01111101010011011	s_4^*	010**0**011**11**	s_4^*	1111111100100
s_5^*	00011101110011011	s_5^*	110**0**110**10**	s_5^*	0001111110100
s_6^*	00011101110111011	s_6^*	101**0**011**00**	s_6^*	1001111110100
s_7^*	00011101100011101	s_7^*	101**0**011**01**	s_7^*	0001101110110
s_8^*	01011101100011101	s_8^*	111**1**010**00**	s_8^*	0101101110110
s_9^*	00011101100011111	s_9^*	010**0**010**10**	s_9^*	1001101110110
s_{10}^*	01011101100011111	s_{10}^*	111**0**010**10**	s_{10}^*	1101101110110

The irrelevant features for the three datasets with multiple global optima are removed and the local optima networks reconstructed. The expectation is that the number of global optima should decrease. Figure 4 shows the newly constructed local optima networks for the three respective datasets when removing the irrelevant features as determined in Table 5.

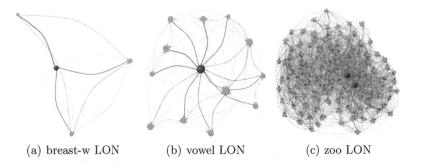

(a) breast-w LON   (b) vowel LON   (c) zoo LON

Fig. 4. Local optima networks after feature removal.

Visually, the local optima networks for the breast-w and vowel datasets indicate the existence of a single global optimum after removal of the irrelevant features. The zoo dataset, however, still has two global optima. Another feature has now been highlighted as being irrelevant, after removal of the first irrelevant feature that was detected in Table 5. This occurrence highlights the complex relationship between features and how the existence of one feature may affect the informational value of other features. The irrelevant feature is determined by calculating the symmetric intersection between the sets of features for the two global optimum solutions. After removal of the irrelevant feature, the local optima network for the zoo dataset is shown in Fig. 5 (now with two features removed). The fitness landscape of the zoo problem now has a single global optimum.

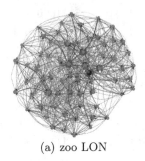

(a) zoo LON

Fig. 5. Local optima networks after feature removal, round 2.

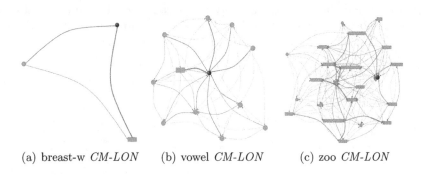

(a) breast-w *CM-LON*   (b) vowel *CM-LON*   (c) zoo *CM-LON*

Fig. 6. Compressed monotonic local optima networks after feature removal

The number of neutral edges in the zoo local optima network after removal of two irrelevant features still stands out among the non-global local optima. The number of neutral edges for the vowel and breast-w datasets after removing the irrelevant features has visually decreased. The compressed monotonic local optima networks are given in Fig. 6, where all irrelevant features have been removed.

The contrast between the compressed monotonic local optima networks for Fig. 6 (after irrelevant feature removal) and Fig. 3 (before irrelevant feature removal) shows the reduction in the number and size of the local optima neutral plateaus. The local optima network metrics for the now single global optima datasets are given in Table 6.

Table 6. Local optima network metrics after feature removal.

Dataset	N_optima	N_global	Bsize_global	Str_global	N_coptima	Prop_neutral
zoo	656	1	0.0039	0.0452	193	0.7057
vowel	13	1	0.1252	0.5948	12	0.0769
breast-w	3	1	0.3513	0.6522	3	0.25

The feature selection problem dimensionality reduction obtained by removing the irrelevant features is apparent in the now reduced number of local optima as indicated in Table 6.

The visual observation of the reduction of neutrality in local optima networks in Fig. 4 is confirmed by the metrics in Table 6. The proportion of connected nodes with neutral edges for the vowel and breast-w datasets decreased drastically from 0.6428 to 0.0769 and 0.65 to 0.25, respectively. The zoo dataset showed a smaller reduction less, from 0.7282 to 0.7057.

The removal of a feature changes the search space of the feature selection problem, which will affect the behaviour of the feature selection search process. The distribution of local optima and the shape of the fitness landscape changes, therefore search algorithms have different information to guide search. Running the feature selection algorithms on the now reduced datasets yield the performance as summarised in Table 7. The values in bold represent an improvement in performance.

Table 7. Feature selection algorithm performance after feature removal.

Dataset	$P_{ratio}(Filter)$	$P_{ratio}(SFS)$	$P_{ratio}(GA)$	GA success ratio
zoo	0.9711	0.8281	**0.9890**	0.1333
vowel	0.8228	1.0	**0.9888**	**0.8**
breast-w	**0.9923**	**0.9923**	**0.9997**	**0.9666**

The genetic algorithm wrapper method performed better for all of the datasets after removal of the irrelevant features, with significant improvement in the success ratio for the vowel and breast-w datasets improving by 14.3%. As a result of the increased success ratio, the performance ratio for the problems also increased, albeit slightly. A Mann-Whitney U test on the samples of performance ratio values for each run of the genetic algorithm, was conducted at a $p = 0.05$ level of significance. The Mann-Whitney U test indicated that the performance improvement is not statistically significant with regards to the performance ratio. The zoo dataset did not show an improvement in success ratio and showed a negligible increase in performance ratio at the third decimal.

The performance of the filter and sequential forward selection algorithms was not affected by the removal of the irrelevant features for the zoo and vowel datasets. A small to negligible improvement in performance ratio is observed on the breast-w dataset for the filter and wrapper methods. The GA wrapper method did not improve in its success ratio, which could be due to other factors affecting the difficulty of the problem. The LON for this problem is observed to contain local optima neutral plateaus that persist after the removal of the neutrality directly attributed to irrelevant features, as seen in Fig. 6.

As previously mentioned, the number of local optima are correlated with the size of the search space but are also affected by the presence of irrelevant features. The zoo dataset, before irrelevant feature removal, exhibits a ratio of 0.0376 ($\frac{4938}{2^{17}}$) local optima with respect to the total number of solutions. After irrelevant feature removal the ratio changes to 0.02 ($\frac{656}{2^{15}}$). The vowel and breast-w ratio of local optima to number of solutions changed from 0.0051 and 0.0781 to 0.0063 and 0.0469, respectively. The removal of the irrelevant features therefore resulted in a reduction in the ratio of local optima to number of solutions by 46.8% for the zoo dataset and 39.9% for the breast-w dataset. The vowel dataset shows an increase in the number of optima at the third decimal which is seen as a negligible change in local optima ratio to number of solutions.

The normalised incoming strength of the global optimum consistently increases for all the datasets. The increased strength to the global optimum from connected local optima indicates that there is a better chance for search to move from a connected local optimum to the global optimum. Removing irrelevant features therefore resulted in a higher probability of reaching a global optimum and can be regarded as a reduction in search difficulty for algorithms utilising neighbourhood in the search space.

5 Conclusion

The purpose of this paper was to conduct an analysis on the fitness landscapes of the feature selection problem.

The construction of local optima networks for the feature selection problem exhibited interesting problem properties that were previously unknown. These insights allowed a better understanding of the performance differences between feature selection algorithms for search spaces with and without irrelevant features.

The construction of the local optima networks for the real world datasets that were used, showed that the feature selection problem had either a single global optimum or several global optima. For problems with multiple global optima, the global optima were located within a single bit-flip neighbourhood of other global optima forming a neutral plateau. Since the inclusion or exclusion of a particular feature had no effect on fitness (classification accuracy), it could be deduced that these features were irrelevant to the classification task.

The incoming strength of escape edges to the global optima represents the probability of moving towards the global optima from a connected local optima. The consistent increase of the normalised incoming strength, after irrelevant feature removal, of the global optimum indicates a reduction in problem difficulty. The proportion of local optima connected via neutral edges in the local optima network consistently reduced when irrelevant features were removed. The proportion of local optima connected via neutral edges fluctuated from a low amount to a very high amount on a per problem basis.

The performance ratio of the filter and wrapper feature selection algorithms did not significantly improve after removing irrelevant features. The success ratio of the genetic algorithm did improve, and remained the same for one dataset, after removing irrelevant features. There was no statistically significant improvement in the performance ratio of the genetic algorithm wrapper method. The improvement in success ratio serves as an indication that the genetic algorithm found the global optima more often and is therefore an indication that the problem difficulty decreased.

Irrelevant features introduce misinformation in the search space, as evident in the reduction of the proportion of local optima that exists in relation to the size of the search space. Removal of irrelevant features generally reduces the number of local optima, both as a result of the dimensionality reduction and removal of misinformation.

The insights gained on the effect of irrelevant features on the fitness landscape and the global structure of the feature selection problem informs on the behaviour of the feature selection algorithms and contributes to the greater goal of meta-learning. Possible future work includes the study of the effect on the fitness landscape using different classifiers and other approaches to feature selection. The effect of increased dimensionality on the fitness landscape is another area of interest for future research.

References

1. Aha, D., Kibler, D.: Instance-based learning algorithms. Mach. Learn. **6**, 37–66 (1991)
2. Amaldi, E., Kann, V.: On the approximability of minimizing nonzero variables or unsatisfied relations in linear systems. Theoret. Comput. Sci. **209**(1), 237–260 (1998)
3. Ben-David, A.: Comparison of classification accuracy using cohen's weighted kappa. Expert Syst. Appl. **34**(2), 825–832 (2008)
4. Chandrashekar, G., Sahin, F.: A survey on feature selection methods. Comput. Electr. Eng. **40**(1), 16–28 (2014)
5. Cover, T.M., Thomas, J.A.: Elements of Information Theory. Wiley, New York (2012)
6. Daolio, F., Verel, S., Ochoa, G., Tomassini, M.: Local optima networks of the quadratic assignment problem. In: 2010 IEEE Congress on Evolutionary Computation (CEC), pp. 1–8. IEEE (2010)

7. Doye, J.P., Massen, C.P.: Characterizing the network topology of the energy landscapes of atomic clusters. J. Chem. Phys. **122**(8), 84105 (2005)

8. Fieldsend, J.E.: Computationally efficient local optima network construction. In: Proceedings of the Genetic and Evolutionary Computation Conference Companion, pp. 1481–1488. ACM (2018)

9. Goldberg, D.E.: Genetic Algorithms in Search, Optimization and Machine Learning. Addison-Wesley, Boston (1989)

10. Guyon, I., Elisseeff, A.: An introduction to variable and feature selection. J. Mach. Learn. Res. **3**, 1157–1182 (2003)

11. Hall, M., Frank, E., Holmes, G., Pfahringer, B., Reutemann, P., Witten, I.H.: The weka data mining software: an update. SIGKDD Explor. Newsl. **11**(1), 10–18 (2009). https://doi.org/10.1145/1656274.1656278

12. Hancer, E., Xue, B., Zhang, M.: Differential evolution for filter feature selection based on information theory and feature ranking. Knowl.-Based Syst. **140**, 103–119 (2018)

13. Kira, K., Rendell, L.A.: The feature selection problem: traditional methods and a new algorithm. In: AAAI, vol. 2, pp. 129–134 (1992)

14. Kohavi, R., John, G.H.: Wrappers for feature subset selection. Artif. Intell. **97**(1), 273–324 (1997)

15. Lichman, M.: UCI machine learning repository (2013). http://archive.ics.uci.edu/ml

16. Malan, K.M., Engelbrecht, A.P.: A survey of techniques for characterising fitness landscapes and some possible ways forward. Inf. Sci. **241**, 148–163 (2013)

17. Ochoa, G., Tomassini, M., Vérel, S., Darabos, C.: A study of NK landscapes' basins and local optima networks. In: Proceedings of the 10th Annual Conference on Genetic and Evolutionary Computation, pp. 555–562. ACM (2008)

18. Ochoa, G., Veerapen, N., Daolio, F., Tomassini, M.: Understanding phase transitions with local optima networks: number partitioning as a case study. In: Hu, B., López-Ibáñez, M. (eds.) EvoCOP 2017. LNCS, vol. 10197, pp. 233–248. Springer, Cham (2017). https://doi.org/10.1007/978-3-319-55453-2_16

19. Ochoa, G., Verel, S., Daolio, F., Tomassini, M.: Local optima networks: a new model of combinatorial fitness landscapes. In: Richter, H., Engelbrecht, A. (eds.) Recent Advances in the Theory and Application of Fitness Landscapes. ECC, vol. 6, pp. 233–262. Springer, Heidelberg (2014). https://doi.org/10.1007/978-3-642-41888-4_9

20. Pitzer, E., Affenzeller, M.: A comprehensive survey on fitness landscape analysis. In: Fodor, J., Klempous, R., Suárez Araujo, C.P. (eds.) Recent Advances in Intelligent Engineering Systems, vol. 378, pp. 161–191. Springer, Heidelberg (2012). https://doi.org/10.1007/978-3-642-23229-9_8

21. Reidys, C.M., Stadler, P.F.: Combinatorial landscapes. SIAM Rev. **44**(1), 3–54 (2002)

22. Vérel, S., Daolio, F., Ochoa, G., Tomassini, M.: Local optima networks with escape edges. In: Hao, J.-K., Legrand, P., Collet, P., Monmarché, N., Lutton, E., Schoenauer, M. (eds.) EA 2011. LNCS, vol. 7401, pp. 49–60. Springer, Heidelberg (2012). https://doi.org/10.1007/978-3-642-35533-2_5

23. Xue, B., Zhang, M., Browne, W.N.: Particle swarm optimization for feature selection in classification: a multi-objective approach. IEEE Trans. Cybern. **43**(6), 1656–1671 (2013)

24. Yu, L., Liu, H.: Feature selection for high-dimensional data: a fast correlation-based filter solution. In: ICML, vol. 3, pp. 856–863 (2003)

Clarifying the Difference in Local Optima Network Sampling Algorithms

Sarah L. Thomson[1]([✉]) [iD], Gabriela Ochoa[1] [iD], and Sébastien Verel[2] [iD]

[1] Computing Science and Mathematics, University of Stirling, Stirling, UK
{s.l.thomson,gabriela.ochoa}@stir.ac.uk
[2] Université du Littoral Côte d'Opale, EA 4491 - LISIC, Calais, France
verel@uni-littoral.fr

Abstract. We conduct the first ever statistical comparison between two *Local Optima Network* (LON) sampling algorithms. These methodologies attempt to capture the connectivity in the local optima space of a fitness landscape. One sampling algorithm is based on a random-walk *snowballing* procedure, while the other is centred around multiple traced runs of an Iterated Local Search. Both of these are proposed for the Quadratic Assignment Problem (QAP), making this the focus of our study. It is important to note the sampling algorithm frameworks could easily be modified for other domains. In our study descriptive statistics for the obtained search space samples are contrasted and commented on. The LON features are also used in linear mixed models and random forest regression for predicting heuristic optimisation performance of two prominent heuristics for the QAP on the underlying combinatorial problems. The model results are then used to make deductions about the sampling algorithms' utility. We also propose a specific set of LON metrics for use in future predictive models alongside previously-proposed network metrics, demonstrating the payoff in doing so.

Keywords: Combinatorial fitness landscapes ·
Local optima networks · Quadratic Assignment Problem

1 Introduction

Local optima networks (*LON*) are a partial fitness landscape of a combinatorial optimisation problem [1]. Features of the networks have repeatedly been linked to heuristic search in various famous problem domains ([2–5]). LONs were originally used for fully-enumerated fitness landscapes and therefore reflected small problems. More recently, sampling algorithms have been proposed [5–7] with the aim of using the analysis on larger problems. The nature, biases and resultant samples of the algorithms are of critical importance in moving LON analysis towards real-world systems. In this study we conduct the first comparison of sampling algorithms for this purpose. Features of the samples are contrasted, and the

The paper "Clarifying the Difference in Local Optima Network Sampling Algorithms" contained an error. The correct text was added at the end of the correction chapter. The correction to this chapter is available at
https://doi.org/10.1007/978-3-030-16711-0_14

A. Liefooghe and L. Paquete (Eds.): EvoCOP 2019, LNCS 11452, pp. 163–178, 2019.
https://doi.org/10.1007/978-3-030-16711-0_11

features are used in linear and random forest models for predicting heuristic performance on the underlying combinatorial problems. In other words, we examine which of the two most recent sampling algorithms ([6,7]) produces LONs with more predictive power, in terms of their ability to measure search difficulty. We also propose using funnel metrics as features for predictive models of algorithm performance. A *funnel* is a large fitness landscape feature and is essentially a basin of attraction at the level of local optima. The main contributions of this paper are:

1. The first descriptive comparison of local optima network sampling algorithms.
2. Contrast of the differences in predictive power of the LON features for explaining meta-heuristic variance.
3. The addition of a specific funnel metric set for predicting algorithm performance in LON models using linear and random forest regression.

2   Definitions

A fitness landscape [8] is a triplet (S, N, f) where S is the set of all possible solutions, $N : S \longrightarrow 2^S$, a neighbourhood structure, is a function that assigns to every $s \in S$ a set of neighbours $N(s)$, and f is a fitness (objective value) function such that $f : S \longrightarrow \mathbb{R}$, where the fitness value is a real number that can be viewed as the *height* of a given solution in the landscape.

A *local optima network* is a representation of the fitness landscape at the level of local optima. We now formally define the constituent parts of a local optima network, before proceeding to describe the object as a whole.

Nodes. The set of nodes, LO, is comprised of local optima, i.e. a solution lo_i satisfies the condition that it has superior fitness to all other solutions in its neighbourhood: $\forall n \in N(lo_i) : f(lo_i) \geq f(n)$, where $N(lo_i)$ is the neighbourhood and n is a single neighbour.

Edges. The set of edges, E, consists of directed and weighted links. An edge is traced if the probability of 'escape'—using perturbation and then hill-climbing—from the source node to the destination is greater than zero, and is weighted with the probability. Formally, local optima lo_i and lo_j form the source and destination of an edge iff $w_{ij} > 0$.

Local Optima Network (LON). The weighted local optima network $LON = (LO, E)$ is a graph where the nodes $lo_i \in LO$ are the local optima, and there exists an edge $e_{ij} \in E$, with weight w_{ij}, between two nodes lo_i and lo_j if $w_{ij} > 0$. Note that w_{ij} may be different than w_{ji}. Thus, two weights are needed in general, and so a local optima network is an oriented transition graph.

2.1   The 'Network' Feature Set

We now describe and introduce terminologies for the complex network features we use in this study. The number of edges found in one of our sampled networks is referred to as *edges*. We also use the mean fitness found in the sample of local optima comprising the network nodes, and denote this as *meanfitness*. The network diameter (the longest possible length in number of edges separating two nodes), *diam*, and the average out-degree (the number of edges which are directed away from a particular node), *outdegree*, are also included.

2.2 The 'Funnel' Feature Set

A fitness landscape *funnel* is a basin of attraction at the level of local optima. Instead of a single local optimum as an attractor for surrounding solutions, every member of a funnel is a local optimum. Each of these is on a path which, if followed, terminates at a single high-quality local optimum—the bottom of the funnel. These funnel bottoms are also referred to as *sinks* from a graph theory perspective, and are simply the nodes with no outgoing edges. We uncover the funnels by first pruning off any LON edges which are non-improving in fitness, and then by commencing a depth-first search from a funnel bottom, in the process revealing the sequences of local optima which must terminate there.

We use four features relating to fitness landscape funnels in our analysis here. The first is simply the amount of funnels as a proportion of the total node count, *funnel*. We also use the number of compressed local optima. The compression process happens during the network pre-processing for funnel analysis, where the non-improving edges are removed, and then plateaus at the local optima level are joined into a single node. The metric *ncoptima* is the number of optima left after this processing of the network. Also considered is the incoming flow to sub-optimal funnel bottoms or sinks (which can be calculated by the addition of the incoming edge-weights to the sink nodes)—this is referred to as *substrength*, and it comes as a proportion of the total flow going to all funnel bottoms, including optimal ones. The fourth funnel feature is *sinkfitness*, which is simply the mean fitness of funnel bottom nodes (sinks in the network).

3 Experimental Setting

3.1 Benchmark Test Problem

Our test instances are Quadratic Assignment Problems (QAP), where an optimal assignment of N facilities to N locations is sought. Each facility, location pair has an associated distance and flow between them. An instance of QAP, then, is defined with two matrices: a *distances* matrix, and a *flow* matrix. A solution to the QAP, π, is a permutation of facilities, while the objective value is calculated as follows:

$$C(\pi) = \sum_{i=1}^{n} \sum_{j=1}^{n} a_{ij} b_{\pi_i \pi_j}$$

where $A = \{a_{ij}\}$ and $B = \{b_{ij}\}$ are the distance and flow matrices, respectively.

We focus exclusively on the much-studied Quadratic Assignment Problem Library (QAPLIB). QAPLIB boasts a diverse selection of problems, both synthetic and real-world, from different classes. Here we use 30 moderately-sized instances, from various problem classes. The problem dimensions N are between 25 and 50, meaning 25–50 facilities to be assigned. Occurrences of the QAP can largely be placed in one of four categories: Uniform random distances and flows; random flows on grids; real-world problems; and random real-world like problems. Instances from all of these categories are considered.

Uniform Random Distances and Flows. This class of QAP is known to be difficult for optimisation heuristics. The entries for both the distance and flow matrices are taken at random from a Gaussian distribution. The naming convention for them in the QAPLIB is *tainn*, with *nn* being the problem dimension. The nature of the points on the plane is random. We use here *tai25a*, *tai30a*, *tai35a*, and *tai40a*.

Random Flows on Grids. For these problems, the locations are each situated in one square on an *m*n* grid, which is rectangular. The flow matrix entries are generated randomly. From this category, we use *nug25*, *nug27*, *nug28*, *nug30*, *sko42*, *tho30*, *tho40*, and *wil50*.

Real-World Problems. These problems arise from practical applications. We briefly describe the instances used in this article. We use two *bur* instances, which comprise of the average stenotypist's typing data. Also included are some *kra* problems: these were used to plan a German hospital. The *esc* instance we use presents itself in computer science (the testing of sequential circuits). In the *ste* problem set, the goal is to minimise the length of connections between units that have to be placed on a rectangular grid. The real-world instances we use in our analysis are *bur26a*, *bur26b*, *esc32e*, *kra30a*, *kra30b*, *kra32*, *lipa30a*, *lipa30b*, *lipa40a*, *lipa40b*, *ste36a*, *ste36b*, and *ste36c*.

Random Real-World like Problems. The naming convention is *tainnb*, where *nn* is the problem dimension. We use the problems *tai25b*, *tai30b*, *tai35b*, and *tai40b*.

Miscellaneous. The *chr* set of problems are a special occurrence of the QAP and do not fit into any of the previous categories. The flow matrix forms a mathematical *tree*, with no particular specification for the definition of the distance matrix. We use *chr25a* from this set.

3.2   The Sampling Algorithms

Snowball (Chain-Referral) Sampling. Snowball sampling is a process, originating in the social sciences, where each respondent in a survey asks a few of their friends to complete the survey as well. The social network of survey respondents is then analogous to an ever-growing rolling snowball, growing larger with every step.

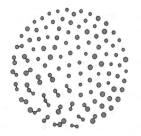 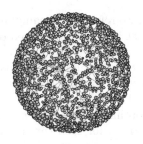

(a) *optSample* LON sample for instance *wil50*. Top 5% of optima by fitness.

(b) *walkSample* LON sample for instance *wil50*. Top 0.5% of optima by fitness.

Fig. 1. A representative LON for an instance from the Quadratic Assignment Problem Library. Figures show LON samples as obtained by two different sampling methods. The global optimum or optima in the samples are shown in red. (Color figure online)

This was introduced as the mechanism behind a sampling algorithm for local optima networks very recently [7].

To construct the local optima network sample, the algorithm first starts from a random solution and hill-climbs to a local optimum, x_t. This local optimum is the start-point of the walk. Then the recursive snowball procedure branches from x_t: the neighbouring local optima are explored, and then the neighbours of those are explored, up to a specified 'depth' given by a parameter d. The local optima are found by carrying out mutations followed by hill-climbing. The hill-climbing used to obtain local optima is best-improvement and uses a random pairwise swap of facilities as the operator. The perturbation operator is four of these mutations in a row, i.e. four swaps.

After the snowball expansion of x_t, the algorithm returns to the initial local optimum x_t and continues on its walk from there. Specifically, a neighbouring local optimum *which is not already present in the walk* is found. If all neighbouring local optima have been expanded, a new random solution is generated and hill-climbing applied to find a local optimum for inclusion in the random walk. Then, that optimum is included in the walk as x_{t+1} and becomes the 'centre' from which a new iteration of snowballing expansion begins. The algorithm then goes back x_{t+1} and finds a neighbour to be x_{t+2} on the walk, and so on. The walk $x_{t...t+n}$ is of a length given by the sampling parameter l.

LON snowball sampling is configurable with three parameters: l (length of random walk), d (depth of snowballing), and m (number of sampled edges). We use the three (l, d) combinations suggested in [7] for obtaining our LONs: $(100, 60)$, $(400, 30)$, and $(100, 30)$, alongside the suggested value for m, which is two. The parameters l and d can be used to tune the sample obtained.

A full description of LON snowballing, including pseudo-code and descriptive figures, can be found in [7] and its Algorithm 2 is particularly helpful. In the text that follows, we refer to this sampling algorithm using the term *walkSample*, to reflect the fact it is a sample based on a random walk.

Iterated Local Search Sampling. This sampling method is built around the competent *Improved Iterated Local Search* [9] heuristic for the QAP, and was proposed in [6]. The local optima network is built while the heuristic algorithm is attempting optimisation. In essence, 200 ILS runs are commenced from different random solutions; for each run, as the ILS progresses in the minimisation process, all local optima encountered are recorded. Alongside these, every transition between two local optima (using perturbation and then hill-climbing) is saved as an edge in the network (LON), or if the edge has already been recorded, the weight of the edge is incremented. Each run terminates either when the global optimum is found, or after 10000 iterations without improvement.

The local search stage uses a first improvement hill-climbing variant with a random pairwise swap operator (the same as the LS in *walkSample*. The perturbation operator exchanges k randomly chosen items. In the original ILS algorithm code, a few settings for k are suggested. We use a subset of those in this study, namely: $\frac{n}{4}, \frac{n}{2}$, and $\frac{3n}{4}$. These settings are important for the construction of the LONs, because the perturbation mechanism controls the discovery and inclusion of network edges (connections between nodes). A comprehensive description of both the original ILS and the modified version for LON sampling can be found in [6]. We refer to this sampling process henceforth by *optSample*, to reflect the fact that the sampling is done alongside heurisitic optimisation.

Figure 1 compares a sample from *optSample* algorithm with one obtained using *walkSample*. The Figure serves as a precursor to a formal comparative analysis later on, but captures a remarkable amount of information. Both are sampled local optima networks, extracted from the same Quadratic Assignment Problem. The sample in Fig. 1a is extracted using *optSample*, while *walkSample* was used to obtain the network in Fig. 1b.

In red is the global optimum (or the global optima), while all other nodes are grey in colour. Both samples have been capped by fitness: only optima in the top 5% (Fig. 1a) and 0.05% (Fig. 1b) of the fitness distributions are included in the plots. The latter percentile is lower because the sample had orders of magnitude more optima and edges crowding its network at this fitness level.

Comparing the two plots, there are striking differences. Taking them in turn, we can see that the high-quality sub-LON which *optSample* produces (Fig. 1a) is extremely sparse, and indeed the two global optima are disconnected (isolated) nodes in this graph. This could indicate that a heuristic would be unlikely to end up at one of these apparently inaccessible nodes; alternatively, it could be that the global optima are well connected nodes (or even network *hubs*), but their network neighbours exist in lower-quality fitness levels.

Casting our attention to the LON in Fig. 1b—produced by *walkSample*— it is a markedly more dense network. There are significantly more nodes and edges between them. Recalling that this is only the highest-quality 0.05% of local optima which were sampled, the implication is that this sampling method exposes much of the neighbourhood (at the local optima level) surrounding the global optimum, and may be more suited to characterising this promising area of the search space. A global optimum can be seen in the centre Fig. 1b.

3.3 Heuristics

To gain a view of empirical complexity of the chosen problem instances, we must collect optimisation data by competent heuristic algorithms.

Two competitive algorithms from the literature are deployed: Stützle's Improved ILS for the QAP (IILS) [9], and Taillard's Robust Taboo Search for the QAP (TS) [10]. IILS comes with a wealth of viable parameter configurations. We choose here first improvement hill-climbing, in combination with a perturbation strength of $\frac{3n}{4}$ pairwise exchanges of facilities.

To quantify their performance on the thirty problems, we use the performance gap metric, which is the obtained fitness as a proportion of the desired fitness after a fixed budget of iterations (1000 here). We take the average of this value over 1000 runs. In the results that follow, this metric is referred to as *IILSp* and *TSp* (shorthand for ILS and TS performance).

3.4 Predictive Model Setup

We use both linear mixed models and random forest regression models for algorithm performance prediction. When considering our full (90 sampled LON per sampling algorithm) sets of observations, we conduct random repeated subsampling cross-validation (also known as bootstrapping) for 1000 iterations, each time shuffling the observations randomly. We do this with an 80–20 training-test split. The predictors are normalised as $p = \frac{(p - E(p))}{sd(p)}$. For the smaller datasets (split by parameter set), we only have 30 observations, so we do not split into training and test data, but use all 30 for regression. We still use bootstrapping to average the model statistics over 1000 iterations.

In the results that follow, we focus on the adjusted R^2 associated with each of the models, which quantifies the amount of variance in the response variable (in our case either *IILSp* or *TSp*) explainable using the set of features in question (either the full metric set, which is the union of the network and funnel features, or the funnel or network features exclusively).

For the random forest regression models, we also report the best four predictors for the models in terms of importance, computed during the model formation. We do not conduct random forest regression on the smaller sets of 30 LONs—only on our two full sets of 90 LONs—to provide enough data for the process.

4   Results

4.1 Sampling Algorithm Comparison

To have two sampling algorithms produce networks which are somehow correlated would be encouraging. We rely on sampling algorithms for local optima networks on problems of any reasonable size. As a result, before the algorithms reach real-world applications, they must be refined. Table 1 provides Spearman correlation coefficients, calculated between features of LONs obtained by *optSample* and *walkSample*, respectively. An indication of the p-value is given, as described in the caption. Both network features and funnel features are included as variables. These were introduced in Sects. 2.1 and 2.2, respectively.

Most of the features have a fairly weak correlation, but with a p-value less than the threshold 0.05, giving evidence against the null hypothesis that the sample features are unrelated. The correlation for the out-degree of the samples is moderate, and has an encouragingly small p-value of < 0.001. The two strongest associations are for the mean fitness in the network, and the mean fitness of the funnel bottoms (*sinkfitness*), respectively. These two features seem to show a great deal of agreement, gaining very strong correlations indicating statistical significance.

Table 2 attempts to quantify the predictability of *optSample* and *walkSample*. This is done by looking at the ranges of values for important features of the obtained LONs. The minimum value in the sample set is represented as a proportion of the maximum value found.

Looking first at the column giving the information on *optSample*, we can see that the proportions are mostly much nearer to zero than to one. This means that the smallest values are completely dwarfed by the values of the largest. The deduction from this is that you cannot tell this sampling methodology exactly what which trajectory to follow. This is advantageous in the sense that it is not rigid or artificially contained, but could also have the potential to result in unreliable samples. It is, however, promising when considering the wide range of the average outdegree of the obtained networks; this hints at hub-and-spoke network structure being exposed (where it exists).

In contrast, the column showing the range information for *walkSample* shows much larger proportions between 0.29 and 0.99, telling us that this algorithm produces a fairly predictable and therefore tuneable number of nodes and edges. However, because the range for the outdegree is very small (the smallest outdegree is almost as big as the largest, arising due to the nature of the algorithm), important connections between local optima will likely be missed and so-called network 'hubs' may never be found. These are critical because much of the flow of the network—in our case, prospective heuristic search trajectories—may pass through them or be drawn towards them.

4.2   Prediction of Heuristic Competence on the Problems

For our predictions, we computed models based on all 90 sampled LONs (that is, 90 per sampling algorithm), which comprise LONs generated using three parameter sets, each of them used for all 30 instances.

We also wanted to see the predictive power when using LONs generated using a single, fixed set of parameters; therefore, we also considered subsets of the LON datasets, where we split into sets of 30.

Table 1. Spearman's rank correlation, $^{***}p < 0.001$, $^{**}p < 0.01$, $^*p < 0.05$, for pairs of features from the two sampling algorithms.

Predictor	Corr.
optima	0.239*
edges	0.149
ncoptima	0.249*
outdegree	0.441***
diam	−0.212*
meanfitness	0.990***
sinks	0.112
sinkfitness	0.990***
substrength	0.274**

Table 2. The variance (represented as the minimum value as a proportion of the maximum value) of some key metrics for the sampled LONs.

Predictor	*optSample*	*walkSample*
optima	0.0003	0.2499
edges	0.0003	0.2499
outdegree	0.0508	0.9999
sinks	0.0003	0.2499

For *optSample*, the parameter difference came from the perturbation setting used in the construction of the edges, connecting optima in the sample. The three settings of this were $\frac{n}{4}, \frac{n}{2}$ and $\frac{3n}{4}$, and the LON sets corresponding to these are referred to as set 1, set 2, and set 3 (in that order) in the models that follow. The set of all 90 LONs together is set 4.

In the case of *walkSample*, the parameter combinations were of the sampling depth, d and length of walk, l. The sets used were (60, 100), (30, 400), and (30, 100); these are referred to as set 1, set 2, and set 3, respectively. Again, the set of all 90 LONs combined is set 4.

Table 3 reports the adjusted R^2 for several predictive models. Each row gives information on the models in question. The columns represent, in order, the sampling algorithm used to obtain the LON samples (*Sample*); the type of predictive methodology (*Model*, one of random forest, *rf*, or linear mixed model, *lm*); the set of features used as predictors (*Features*); the parameter set used in the construction of the network samples (*Param*); and an indication of whether *ILS* or *TS* performance gap was the response variable. The higher the values for the models, the more explainable the variance in the optimisation metric is using that set of predictors.

Table 3. Adjusted R^2 values for linear and random forest models to explain heuristic performance variation on the combinatorial problems.

Sample	Model	Features	Param.	ILS	TS
optSample	lm	all	set 1	0.733	0.425
optSample	lm	all	set 2	0.727	0.592
optSample	lm	all	set 3	0.757	0.734
optSample	lm	all	set 4	0.673	0.539
optSample	lm	funnel	set 1	0.544	0.144
optSample	lm	funnel	set 2	0.428	0.170
optSample	lm	funnel	set 3	0.487	0.265
optSample	lm	funnel	set 4	0.434	0.167
optSample	lm	network	set 1	0.480	0.399
optSample	lm	network	set 2	0.471	0.424
optSample	lm	network	set 3	0.535	0.595
optSample	lm	network	set 4	0.539	0.366
optSample	rf	all	set 4	0.799	0.543
optSample	rf	funnel	set 4	0.615	0.444
optSample	rf	network	set 4	0.678	0.521
walkSample	lm	all	set 1	0.584	0.299
walkSample	lm	all	set 2	0.723	0.809
walkSample	lm	all	set 3	0.673	0.608
walkSample	lm	all	set 4	0.511	0.328
walkSample	lm	funnel	set 1	0.449	0.289
walkSample	lm	funnel	set 2	0.431	0.276
walkSample	lm	funnel	set 3	0.434	0.393
walkSample	lm	funnel	set 4	0.320	0.328
walkSample	lm	network	set 1	0.355	0.086
walkSample	lm	network	set 2	0.445	0.463
walkSample	lm	network	set 3	0.366	0.136
walkSample	lm	network	set 4	0.262	0.098
walkSample	rf	all	set 4	0.497	0.337
walkSample	rf	funnel	set 4	0.274	0.217
walkSample	rf	network	set 4	0.217	0.001

Focusing on the *optSample* entries, we can see that the general trend is that more variance is explained in the ILS than in the TS (compare *ILS* column with *TS* column). When considering the 90 LONs together, using all the features together (*all*) produces the strongest models of the three predictor sets (*all*, *funnel*, and *network*). These are followed by the network predictor set models,

Table 4. Predictor rankings for the size random forest models.

Sample	Features	Param	Resp.	1	2	3	4
optSample	funnel	set 4	ILS	sinkfitness	ncoptima	funnel	substrength
optSample	network	set 4	ILS	meanfitness	edges	outdegree	diam
optSample	all	set 4	ILS	meanfitness	**sinkfitness**	ncoptima	edges
walkSample	funnel	set 4	ILS	sinkfitness	ncoptima	funnel	substrength
walkSample	network	set 4	ILS	meanfitness	outdegree	edges	diam
walkSample	all	set 4	ILS	meanfitness	**sinkfitness**	outdegree	substrength
optSample	funnel	set 4	TS	sinkfitness	ncoptima	funnel	substrength
optSample	network	set 4	TS	meanfitness	outdegree	edges	diam
optSample	all	set 4	TS	outdegree	**sinkfitness**	edges	meanfitness
walkSample	funnel	set 4	TS	sinkfitness	ncoptima	funnel	substrength
walkSample	network	set 4	TS	meanfitness	outdegree	edges	diam
walkSample	all	set 4	TS	**sinkfitness**	meanfitness	outdegree	ncoptima

with the funnel set coming in lowest. This being said, the R^2 values for the full predictor set are higher than for using the network metrics alone, meaning that including the funnel features provides some extra information on the performance of the ILS and TS.

The poorest fits of the *optSample* linear mixed models come from the prediction of TS response using only funnel variables, with only a maximum of 27% of variance being explained. If we compare the random forest models with the linear mixed models—taking the bottom three *optSample* rows against the set 4 (full set) *optSample* rows—it can be seen that these are stronger than the linear predictions.

The *walkSample* models are summarised in the lower half of Table 3. Let us first consider the values obtained using the full set of 90 sampled LONs (the rows where column *Param.* is stated as *set 4*). Comparing these three with one another, we see that for ILS, the best predictor set is all the metrics together. The funnel set work the next-best, with the network set coming in last for predictive power. For Tabu Search prediction, using the full feature set produces an R^2 value of 0.328. Interestingly, using *exclusively* the funnel set produces the same strength of model. Using only the network features has a very poor result, with only 0.098% of variance in performance gap being explained by these.

Just like with the *optSample* models, the trend over the models here is that more of the variance in ILS is explainable with the predictors than that of the TS. This can be seen by locating the *ILS* column and checking against the neighbouring *TS* column entries.

The random forest models for *walkSample* LONs are generally a bit weaker than their linear counterparts (the bottom three rows compared with the *set 4* rows above). In the random forest section, we can see that for these, using the full set of features produces the highest R^2 values. For ILS, using the funnel or network predictor sets is basically comparable (0.274 and 0.217 respectively), but for TS the network set is basically useless (0.001) while the funnel set has some utility (roughly 22% of variance explained).

We now compare the random forest models for the two sampling methodologies and look at six rows in total (all rows where *Model* is *rf*). What is clear, if going by these alone, is that *optSample* has more predictive power for these problems and algorithms than *walkSample*. This can be seen in the uniformly higher R^2 entries.

Doing the same comparison for the linear models for each of our two sampling algorithms (every row where *Model* is stated as *lm*), it is less of a clear-cut distinction. Granted, the general trend is that *optSample* has higher R^2 values, but there are exceptions, notably in the funnel set predictions for Tabu Search, which are poorer than the equivalent models for the *walkSample*.

Table 4 reports a ranking of features in terms of importance in their predictive random forest models. They are ranked from 1–4. The column labelled '1' shows the most important feature from the model specified in that row. In the cases where the full feature set are used as predictors (the rows where *Features* are stated as 'all'), there were eight variables used, but only the best four are reported in the Table.

We can see from examining the '1' row that almost always, the *meanfitness* (mean local optimum fitness in the network) and *sinkfitness* (mean funnel bottom fitness in the network) are ranked as the most important. *Outdegree* seems to be important for prediction of the response variables too.

Looking to the last column, we can see that *substrength* and *diam* populate most of it, indicating these are of lower importance as predictors.

Correlation Study. Figure 2 is a correlation matrix contrasting observed features (*edges, meanfitness, outdegree, diam, substrength, sinkfitness, funnel*, and *ncoptima*), of the local optima networks sampled by *optSample* with heuristic performance on the underlying combinatorial problem (*IILSp* and *TSp*).

Each entry in the upper-right triangle is a Spearman correlation between the row and column variables. The black text is the overall correlation for all 90 LONs, while the coloured text is the correlations for only LONs from a particular parameter set (red is set 1, green is set 2, and blue is set 3). The diagonal shows density plots for the features, with the lower-left triangle showing pairwise scatterplots.

We are, of course, principally interested in the connections between our heuristic performance metrics, and the features of the sampled LONs. The easiest way to approach this is by locating the *IILSp* and *TSp* rows and looking along them, checking against the intersections for interesting correlations. For *IILSp*, we can see that (if we are taking $p < 0.05$ to be indication of statistical significance), many correlations seem notable: *ncoptima, funnel, sinkfitness, substrength, diam, outdegree, meanfitness*, and *edges*.

The deduction while reading these is that of those listed, the ones with negative coefficients are correlated to a better heuristic search (a lower performance

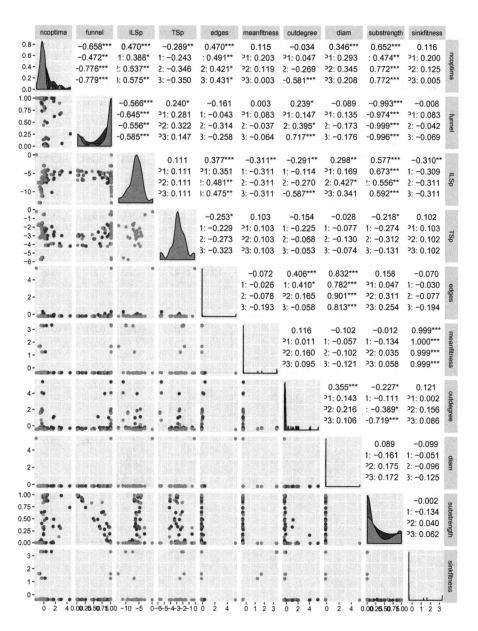

Fig. 2. Correlation matrices of performance metrics and *optSample*-produced LON features. Lower triangle: pairwise scatter plots. Diagonal: density plots. Upper triangle: pairwise Spearman's rank correlation, ***$p < 0.001$, **$p < 0.01$, *$p < 0.05$.

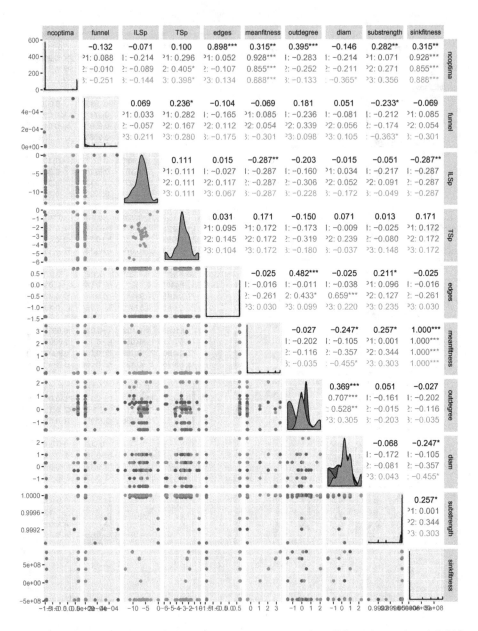

Fig. 3. Correlation matrices of performance metrics and *walkSample*-produced LON features. Lower triangle: pairwise scatter plots. Diagonal: density plots. Upper triangle: pairwise Spearman's rank correlation, $^{***}p < 0.001$, $^{**}p < 0.01$, $^{*}p < 0.05$.

gap), while those with positive values are correlated with a larger performance gap by our heuristics.

For *IILSp*, those that fall into the first (desirable) category are *funnel, sink-fitness, outdegree,* and *meanfitness*. Those correlated with a larger performance gap are *ncoptima, substrength, diam,* and *edges*. In general, the correlations with the features and the performance metrics are stronger in the case of ILS than for TS. However, there are a few to note for TS, selected for their associated p-value: *ncoptima, funnel, substrength,* and *edges*.

Figure 3 has precisely the same layout as Fig. 2, but shows results for *walk-Sample* instead of *optSample*.

If we assess things in the same way: the intersections of *IILSp* and *TSp* with other columns and rows. In general, there are much weaker correlations with the performance gap data here than we saw for the features of the *optSample* LONs. This can be seen in the smaller values and lack of p-value indications for the feature-performance pairs.

There are a few correlations of note, however: for *IILSp*, there are negative correlations with statistically-sound p-values with *meanfitness* and *sinkfitness* – i.e. if optima, and funnel bottoms, are of a higher-quality fitness, this is correlated with a smaller performance gap attained by the ILS. There is also a positive correlation between *TSp* and *funnel*, indicating funnels are correlated with a larger performance gap for TS.

5 Conclusions and Thoughts

We have conducted an empirical comparison between two sampling algorithms for local optima networks of the Quadratic Assignment Problem. It is essential to refine sampling for local optima networks before we bring them to real world systems. A descriptive statistical comparison was reported, as well as findings pertaining to the algorithms' predictive power in estimating the performance gap obtained by two prominent meta-heuristics on the underlying combinatorial problem.

We found that the two sampling methods exhibited some agreement in the networks they produced and that we could reject the null hypothesis that they produce completely independent samples. They differed from a descriptive perspective in that *walkSample* was tuneable and predictable, while *optSample* varied widely but seemed good at finding hub-and-spoke structure in the local optima space.

A correlation study for the LON features and the heuristic performance metrics was conducted. The correlations were stronger and clear when considering the features of the LONs obtained using *optSample* than *walkSample*. We also worked on predicting heuristic algorithm performance on the problems using linear and random forest models, and found that the sampled LON features (for both *optSample* and *walkSample*) better fit the ILS response variable than the TS one. We saw that generally, including both the *funnel* metric set and the *network* set would be advantageous in explaining search discrepancies for these two heuristics. For both *optSample* and *walkSample*, the extracted funnel metrics proved useful. Going off the random forest models alone, *optSample* uniformly had more predictive power than its competitor, for these choices of instances and heuristics.

From the random forest rankings, the most important predictors were those pertaining to fitness in the sampled networks: the fitness of funnel bottoms, and of nodes in general in the network. This hints that perhaps fitness levels in the local optima space are more pertinent to heuristic search than the subset of transition edges sampled by the LON algorithms.

Acknowledgements. This work is supported by the UK's Engineering and Physical Sciences Research Council (grant number EP/J017515/1). Data generated during this research are available from the Stirling Online Repository for Research Data (http://hdl.handle.net/11667/91).

References

1. Ochoa, G., Tomassini, M., Vérel, S., Darabos, C.: A study of NK landscapes' basins and local optima networks. In: Proceedings of the 10th Annual Conference on Genetic and Evolutionary Computation, pp. 555–562. ACM (2008)
2. Ochoa, G., Veerapen, N.: Mapping the global structure of TSP fitness landscapes. J. Heuristics **24**(3), 1–30 (2017)
3. Herrmann, S., Ochoa, G., Rothlauf, F.: Communities of local optima as funnels in fitness landscapes. In: Proceedings of the 2014 Annual Conference on Genetic and Evolutionary Computation, pp. 325–331 (2016)
4. McMenemy, P., Veerapen, N., Ochoa, G.: How perturbation strength shapes the global structure of TSP fitness landscapes. In: European Conference on Evolutionary Computation in Combinatorial Optimization, pp. 34–49 (2018)
5. Iclanzan, D., Daolio, F., Tomassini, M.: Data-driven local optima network characterization of QAPLIB instances. In: Proceedings of the 2014 Annual Conference on Genetic and Evolutionary Computation GECCO 2014, pp. 453–460. ACM, New York (2014). http://doi.acm.org/10.1145/2576768.2598275
6. Ochoa, G., Herrmann, S.: Perturbation strength and the global structure of QAP fitness landscapes. In: Auger, A., Fonseca, C.M., Lourenço, N., Machado, P., Paquete, L., Whitley, D. (eds.) PPSN 2018. LNCS, vol. 11102, pp. 245–256. Springer, Cham (2018). https://doi.org/10.1007/978-3-319-99259-4_20
7. Verel, S., Daolio, F., Ochoa, G., Tomassini, M.: Sampling local optima networks of large combinatorial search spaces: the QAP case. In: Auger, A., Fonseca, C.M., Lourenço, N., Machado, P., Paquete, L., Whitley, D. (eds.) PPSN 2018. LNCS, vol. 11102, pp. 257–268. Springer, Cham (2018). https://doi.org/10.1007/978-3-319-99259-4_21
8. Stadler, P.F.: Fitness landscapes. In: Lässig, M., Valleriani, A. (eds.) Biological Evolution and Statistical Physics, vol. 585, pp. 183–204. Springer, Heidelberg (2002). https://doi.org/10.1007/3-540-45692-9_10
9. Stützle, T.: Iterated local search for the quadratic assignment problem. Eur. J. Oper. Res. **174**(3), 1519–1539 (2006)
10. Taillard, É.: Robust taboo search for the quadratic assignment problem. Parallel Comput. **17**(4–5), 443–455 (1991)

A Unifying View on Recombination Spaces and Abstract Convex Evolutionary Search

Marcos Diez García[✉][ID] and Alberto Moraglio

Department of Computer Science, University of Exeter, Exeter EX4 4QF, UK
{md518,a.moraglio}@exeter.ac.uk

Abstract. Previous work proposed to unify an algebraic theory of fitness landscapes and a geometric framework of evolutionary algorithms (EAs). One of the main goals behind this unification is to develop an analytical method that verifies if a problem's landscape belongs to certain abstract convex landscape classes, where certain recombination-based EAs (without mutation) have polynomial runtime performance. This paper advances such unification by showing that: (a) crossovers can be formally classified according to geometric or algebraic axiomatic properties; and (b) the population behaviour induced by certain crossovers in recombination-based EAs can be formalised in the geometric and algebraic theories. These results make a significant contribution to the basis of an integrated geometric-algebraic framework with which analyse recombination spaces and recombination based EAs.

Keywords: Abstract convex landscape · Abstract convex search ·
Convex hull closure · Geometric crossover · Recombination P-structure

1 Introduction

An open challenge in evolutionary computing is to identify (without running any experiments) if a fitness landscape class is well matched with an evolutionary algorithm (EA) class, where good performance guarantees can be provided. Previous work [1] laid the foundations for an analytical fitness landscape method that determines if a problem's landscape belongs to certain abstract convex landscape classes, where certain recombination-based EAs exhibit polynomial runtime [2]. The theories we began unifying, namely a general geometric framework (GF) of EAs [3] and an algebraic theory of fitness landscapes known as elementary landscapes (EL) [4], complement each other towards such challenge even if they originally pursue distinct aims.

On the one hand, GF naturally frames (geometric) mutation and recombination operators within the same search space structure when viewed as a metric space. Among the many benefits of this approach we find: having a clear way

The full version of this article, including some omitted proofs, can be found online at
https://www.researchgate.net/profile/Marcos_Diez_Garcia.

© Springer Nature Switzerland AG 2019
A. Liefooghe and L. Paquete (Eds.): EvoCOP 2019, LNCS 11452, pp. 179–195, 2019.
https://doi.org/10.1007/978-3-030-16711-0_12

of comparing mutation and crossover in a fitness landscape, the formal design and analysis of EAs across many representations (real vectors, sets, permutations, syntactic trees, etc.), and generalisations of particle swarm optimisation and genetic programming with important applications [3]. Besides, there has been a recent development of general runtime analysis in GF [2]. Particularly, geometric-crossover EAs (without mutation) are proved to find optimal solutions in polynomial time by restricting to classes of problems with certain abstract convex landscapes. However, GF lacks analytic tools to tell whether a given landscape is abstractly convex.

On the other hand, EL provides algebraic means to analyse combinatorial landscapes. Some of them, called 'elementary', have received special attention because: they satisfy desirable conditions related to smoothness (i.e. correlation between fitness values) and structure of local optima, and they are present in important NP-complete problems (e.g. travelling salesman, graph colouring, weight partitioning, etc.). However, the algebraic structures known as P-structures that model crossover are not so readily intuitive as geometric crossovers in GF [5]. Furthermore, studies on population-based search algorithms using P-structures are scarce [6], and no runtime analysis has been done in this algebraic context.

Recombination landscapes are not essentially different in both theories and can be unified, meaning that EL can be applied consistently to analyse certain landscapes arising in GF [1]. This paper extends their unification, by showing that the algebraic approach of EL admits dual formalisations of two fundamental aspects about crossovers and their behaviour in GF: classification of geometric crossovers and abstract convex search (accomplished by geometric-crossover EAs). The main contributions are:

1. Proving the existence of two broad classes of crossovers: those which are recombination P-structures, including geometric crossovers [1], and those which are neither geometric nor recombination P-structures. This justifies that a unification of GF and EL would not be an 'empty' theory (i.e. a theory encompassing all recombination operators).
2. A formalisation of abstract convex search in EL, describing the population-behaviour of recombination P-structure EAs; a class that is larger than complete geometric-crossover EAs. It also provides a characterisation of population-based EAs less restrictive than the existing one in EL.

This paper is organised as follows. Section 2 introduces recombination and abstract convex search in GF. Section 3 introduces recombination in EL and other concepts necessary to unify abstract convex search. Section 4 presents the main results. Section 5 summarises this work and suggests future research.

2   Recombination in the Geometric Framework

Moraglio proposed a general geometric framework (GF) of EAs, independent of the problem and representation of solutions, solely based on an axiomatic definition of distance across metric spaces [3]. In GF, the search space of configurations (representing candidate solutions) is defined as a metric space (X, d), with configuration set X and metric $d : X \times X \to [0, \infty)$. Thus, a metric space constitutes the structure of a fitness landscape (X, d, f) with an arbitrary real-valued fitness function $f : X \to \mathbb{R}$. This geometric view of the search space allows mutation[1] and crossover operators to be conceived within the same space structure, by means of metric balls and metric segments respectively.

Definition 1 (Metric ball and segment [7]). Let (X, d) be any metric space. A *closed ball* centered at point $x \in X$ with radius $r \in \mathbb{R}_{\geq 0}$ is defined as $B_d[x; r] := \{y \in X \mid d(x, y) \leq r\}$. A *geodesic interval* or *metric segment* is defined as $[x; y]_d := \{z \in X \mid d(x, z) + d(z, y) = d(x, y)\}$, where $x, y \in X$ are called the *extremes* of the segment and $d(x, y)$ its *length*.

Many kinds of crossovers can be modelled in GF (e.g. multi-parental or probabilistic). For notational simplicity, we assume crossovers taking two parents and returning one offspring; and, denote by $\text{Im}[\xi(\cdot, \cdot)]$ the set of all offspring produced with non-zero probability by a crossover ξ, called the *image set*.

Definition 2 ([Complete] geometric crossover [3]). Let (X, d) be any metric space. A crossover operator ξ is a *geometric crossover* for (X, d), if all offspring belong to the parents' metric segment. That is, $\forall x, y \in X : \text{Im}[\xi(x, y)] \subseteq [x; y]_d$. If $\forall x, y \in X : \text{Im}[\xi(x, y)] = [x; y]_d$, then it is called *complete geometric crossover*.

Example 1 (Uniform geometric crossover). Let a Hamming metric space $(\mathcal{H}_q^n, d_{\text{H}})$ be defined by a set $\mathcal{H}_q^n := \{0, 1, \ldots, q-1\}^n$ of n-dimensional Hamming sequences on a q-ary alphabet for $q \geq 2$, and the Hamming metric d_{H}. Traditional uniform crossover, denoted $\text{UNIFORM}(x, y)$ for any parents $x, y \in \mathcal{H}_q^n$, is an example of a crossover that is geometric and also complete. For instance, consider parents $100, 001 \in \mathcal{H}_2^3$. Then, we have $\text{Im}[\text{UNIFORM}(100, 001)] = [100; 001]_{d_{\text{H}}}$. The same holds for other pair of parents as well [3].

Example 2 (Non-complete geometric crossover). Consider the traditional one-point crossover on Hamming sequences, but returning only a single offspring with its 'head' coming from the first parent and its 'tail' from the second parent: $\text{ONEPOINT}(i) : \mathcal{H}_q^n \times \mathcal{H}_q^n \to \mathcal{H}_q^n, (x, y) \mapsto z = (x_1, \ldots, x_{i-1}, y_i, \ldots, y_n)$; where z is offspring of parents x and y, at crossover point $i : 1 \leq i \leq n$. Notice for example, $\text{ONEPOINT}(2)(000, 111) = 011$ is not generated by $\text{ONEPOINT}(i)(111, 000)$ for any $1 \leq i \leq 3$. Therefore, it is not a complete geometric crossover, because it is asymmetric, and still is geometric since offspring belong to parents segment [3].

However, some crossovers are not geometric. That is, for any metric on a given set X, there always exist some offspring that lie outside the metric segment of some parents. See Sect. 4.1 for examples of non-geometric crossovers.

---

[1] Geometric mutation is defined by requiring that the mutated offspring belongs to the d-metric ball of its single parent.

Proposition 1 (Existence of non-geometric crossovers [3,8]**).** The set of non-geometric crossovers $\overline{\mathcal{GX}}$ is not empty. Therefore, recombination operators split into two proper classes: geometric $\mathcal{GX}$ and non-geometric $\overline{\mathcal{GX}}$. The class of complete geometric crossovers, $\mathcal{GX}$-complete, is a proper subclass of $\mathcal{GX}$.

Achieving generality in a theory may come at the expense of depth, only being able to assert trivial or tautological statements. Proposition 1 justifies that this is not the case in GF, because it is not a theory of all crossovers, rather it restricts to geometric crossovers only. Moreover, this crossover class is relevant, since many crossovers across different representations are geometric, and polynomial runtime guarantees are possible for geometric-crossover EAs in problem classes with certain abstract convex fitness landscapes [2,3]. Essentially, the abstract convex nature of landscapes is what enables good performance of geometric-crossover EAs.

2.1   Abstract Convexity

To formalise the abstract convex search performed by geometric-crossover EAs (Sect. 2.2), a more general notion of convexity than that of traditional Euclidean spaces is required. A natural choice would be geodesic convexity, since GF understands recombination in terms of geodesic intervals.

Definition 3 (Geodesic convex space [7]**).** Let (X, d) be a metric space. A point $z \in X$ is *geodesically between* two points $x, y \in X$, if and only if $z \in [x; y]_d$. Then, a subset $C \subseteq X$ is *geodesically convex* if and only if all points geodesically between any two points x and y in C are also in C, that is $\forall x, y \in C : [x; y]_d \subseteq C$. The family $\mathcal{C}$ of all geodesically convex subsets C on X is called the *geodesic convexity* of (X, d), and the pair $(X, \mathcal{C})$ a *geodesically convex space*.

Geodesic convexity is not equivalent to convexity (Definition 4); the latter being more abstract, where metrics are undefined and the concepts of neighbourhood or nearness are stated in terms of (convex hull) closure operators in a set-theoretical fashion. We explain next why geodesic convexity is problematic in our context.

Definition 4 (Convexity space [7]**).** A family $\mathcal{C}$ of subsets of a set X is called a *convexity* on X if: (I) the empty set $\emptyset$ and the universal set X are in $\mathcal{C}$; (II) $\mathcal{C}$ is stable under arbitrary intersections, that is if $\mathcal{D} \subseteq \mathcal{C}$, then $\bigcap \mathcal{D} \in \mathcal{C}$; and, (III) $\mathcal{C}$ is stable under arbitrary nested unions, that is if $\mathcal{D} \subseteq \mathcal{C}$ is totally ordered by inclusion, then $\bigcup \mathcal{D} \in \mathcal{C}$. The pair $(X, \mathcal{C})$ is called *convexity space*; and the members of $\mathcal{C}$ are called *convex sets*.

Definition 5 (Convex hull closure [7]**).** Let $(X, \mathcal{C})$ be a convexity space. The *convex hull closure* of a set A, $co(A) := \bigcap \{C \mid A \subseteq C \in \mathcal{C}\}$, is the intersection of all convex supersets $C \in \mathcal{C}$, or equally the smallest convex set, that includes A.

Abstract convex search relies on convexity of sets, and precisely the problem with geodesic convexity is that it cannot guarantee across representations (i.e. metric spaces) that a population set would be a convex set as in Definition 4. This is because metric segments do not share the convexity properties of convex hulls. In other words, metric segments need not be convex hulls: $[x; y]_d \subseteq co(\{x, y\})$ for all $x, y \in X$ [7]. Convexity spaces where equality holds (i.e. geodesic convex sets equal the convex hull of their extreme points) are called *convex geometries* [9].

2.2   Abstract Convex Search

GF postulates that the core behaviour of (recombination-based) EAs can be described axiomatically across representations and problems, and, ultimately, the differences between EAs stem from the representation of solutions and adequacy of search operators [2,3,10]. To support this claim, GF proved that the main operations of geometric-crossover EAs (i.e. selection, geometric crossover and replacement) produce new individuals that always belong to the convex hull of their ancestors; regardless of the representation, problem, geometric crossover, offspring probability distribution imposed on the geometric crossover, selection and replacement strategies [3,10]. This relationship between parents and offspring via convex hulls is the key property characterising the population behaviour of any geometric-crossover EA through generations (Fig. 1).

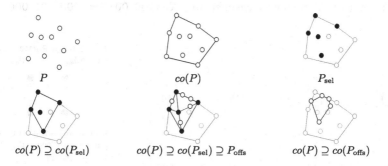

Fig. 1. Abstract convex search in two-dimensional Euclidean space: relationship between parent P, mating pool after selection P_{sel} and offspring P_{offs} populations, and their respective convex hulls $co(P)$, $co(P_{sel})$ and $co(P_{offs})$.

Proposition 2 (Abstract convex search [3]). Let (X, d) be a metric space and $P_t \subseteq X$ a population at generation t. For any EA doing the cycle selection-geometric crossover-replacement: $co(P_0) \supseteq co(P_1) \supseteq \cdots \supseteq co(P_t) \supseteq co(P_{t+1})$.

3   Recombination in Elementary Landscapes Theory

Stadler proposed an algebraic theory of fitness landscapes known as elementary landscapes (EL) [4]. If the geometric approach of GF (Sect. 2) allows us to seamlessly understand mutation and recombination structures in the same space, in EL the situation is more complicated. Originally, EL defines the search space structure $(X, \mathcal{N})$ in terms of (mutation) neighbourhoods $\mathcal{N} : X \to X^k$ mapping an individual to some $k > 0$ possible neighbours (e.g. bit-flip mutation). This naturally leads to formalise the search space as a graph, with the structure given by the adjacency matrix. To model recombination, EL uses an algebraic notion of interval (Definition 6) that captures general recombination structures, not just genetic crossovers [11,12].

Definition 6 (Recombination P-structure [5]). Let X be a non-empty finite set of configurations, $\mathcal{P}(X)$ its power set and an operator $\mathcal{R} : X \times X \to \mathcal{P}(X)$ mapping a pair of parents into a set of possible offspring. We call $(X, \mathcal{R})$ a *recombination P-structure* if $\forall x, y, z \in X$: (I) fix-point: $\mathcal{R}(x, x) = \{x\}$; (II) symmetry: $\mathcal{R}(x, y) = \mathcal{R}(y, x)$; (III) null-recombination: $\{x, y\} \subseteq \mathcal{R}(x, y)$; and (IV) size-monotonicity: if z is in $\mathcal{R}(x, y)$, then $|\mathcal{R}(x, z)| \leq |\mathcal{R}(x, y)|$.

Example 3 (Uniform recombination P-structure). Let parents $x, y \in \mathcal{H}_q^n$ be arbitrary Hamming sequences. Then, $\mathcal{R}_\Omega(x, y) := \{(z_1, \ldots, z_n) \mid z_i \in \{x_i, y_i\}, 1 \leq i \leq n\}$ is a recombination P-structure [11]. For instance, consider $x := 100$ and $y := 001$ in $\mathcal{H}_2^3$, then we have offspring: $\mathcal{R}_\Omega(100, 001) = \{100, 001, 000, 101\}$.

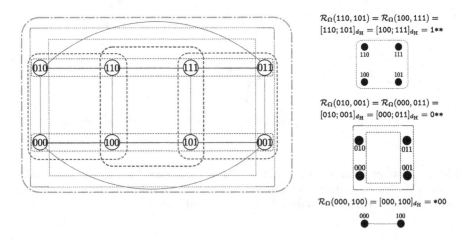

Fig. 2. Hypergraph of the uniform recombination P-structure $\mathcal{R}_\Omega$ in three-dimensional binary Hamming space: vertices are 'circles', hyperedges are both 'solid' and 'dotted-enclosing lines' (left). Enclosed vertices define hyperedges, not the 'lines shapes' or their colours. Examples of specific hyperedges, offspring of $\mathcal{R}_\Omega$ and uniform geometric crossover, and corresponding schema (right).

Example 4 (One-point recombination P-structure). Let parents $x, y \in \mathcal{H}_q^n$ be arbitrary Hamming sequences. Recall the single-offspring one-point crossover $\text{ONEPOINT}(i) : \mathcal{H}_q^n \times \mathcal{H}_q^n \to \mathcal{H}_q^n$, $(x, y) \mapsto z = (x_1, \dots, x_{i-1}, y_i, \dots, y_n)$, $1 \leq i \leq n$, from Example 2. Then, the one-point recombination P-structure is $\mathcal{R}_1(x, y) := \bigcup_{1 \leq i \leq n} \{\text{ONEPOINT}(i)(x, y), \text{ONEPOINT}(i)(y, x)\}$ [12]. For example, $\mathcal{R}_1(000, 111) = \bigcup_{1 \leq i \leq 3} \{\text{ONEPOINT}(i)(000, 111), \text{ONEPOINT}(i)(111, 000)\} = \{0, 1\}^3 \setminus \{010, 101\}$.

However, recombination, acting on pairs of parents, does not have a natural interpretation in graphs. Instead, [5,12] propose hypergraphs (Definition 7), where vertices remain as in graphs but edges are generalised to represent the offspring sets produced by recombining pairs of parents. Figure 2 shows an example for three-dimensional binary sequences and the uniform recombination P-structure $\mathcal{R}_\Omega$, other cases are difficult to visualise clearly.

Figure 2 also illustrates via hyperedges of $\mathcal{R}_\Omega$ hypergraph the following coincidence: offspring produced by the uniform recombination P-structure $\mathcal{R}_\Omega$ and Hamming segments are equivalent [11], and, independently, [10,13] show it for Hamming segments, convex hulls and traditional schema [14]. For instance, the schema $0**$ corresponding to the hyperedge $\{000, 001, 010, 011\}$ is obtained by $\mathcal{R}_\Omega(010, 001)$, $[010; 001]_{d_H}$ or $co(\{010, 001\})$.

Definition 7 (Hypergraph [5]). Let $(X, \mathcal{R})$ be a recombination P-structure. Its *hypergraph* $(X, \mathscr{E})$ has vertex set X and hyperedge set $\mathscr{E} := \{\mathcal{R}(x, y) \mid \mathcal{R}(x, y) \neq \emptyset \wedge x, y \in X\}$, where offspring $\mathcal{R}(x, y)$ represent a hyperedge.

Despite recombination search spaces are then formalised using hypergraphs, they are comparable (i.e. topologically equivalent) to mutation spaces on graphs under certain conditions [5].

3.1   Crossover Random Walk

Discrete-time dynamical systems may be used to model the behaviour of recombination P-structure EAs, but this is often analytically impractical for realistic EAs [6]. Instead, the model adopted in EL is a crossover random walk, a variation on the simple random walk based on mutation moves: start with a prescribed 'father' x_0 in a population, mate it with a 'mother' y_0 randomly sampled from the population to produce a 'child' z_1, then repeat the cycle using the child as father in the next recombination (Fig. 3) [4–6].

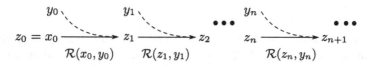

Fig. 3. Schematic representation of a crossover-walk over time.

In this manner, the sequence of fathers (or children) is a crossover random walk. The main reason [4–6] use this model is that the generalised adjacency matrices of hypergraphs, associated to recombination P-structures, can be seen as transition matrices of a Markov process. That is the population search behaviour induced by recombination P-structures can be studied in terms of Markov processes. However, [6] makes two important assumptions that cause the model not to be general. First, it is assumed that there exists none or little statistical interdependence between genes (i.e. epistasis), and low selection genetic pressure; otherwise, the analysis of population behaviour is impractical. Secondly, it is formalised exclusively for traditional string crossovers (e.g. one-point or uniform), not all crossovers that can be described by general recombination P-structures. Generalising this model remains as an open problem.

3.2 Bridge to the Geometric Framework

Figure 2 hints at a close relationship between geometric crossovers and recombination P-structures. This is not coincidental. Indeed, complete geometric crossovers are also recombination P-structures (Proposition 3). Note it does not hold for general geometric crossovers, as [1] unwittingly claimed. For example, any geometric crossover where offspring never include the parents will not fulfil the null-recombination axiom (Definition 6), also the crossover in Example 2 is geometric but not a recombination P-structure due to asymmetry.

Proposition 3 ($\mathcal{GX}$-complete crossovers are recomb. P-structures [1]).
Let (X, d) be any metric space with finite set X, and any crossover $\xi(x, y) := [x; y]_d$ for arbitrary parents $x, y \in X$. Then, $\forall x, y \in X$ and any metric d, (X, ξ) is a recombination P-structure.

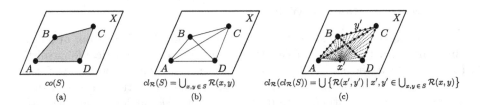

$$co(S) \qquad cl_\mathcal{R}(S) = \bigcup_{x,y \in S} \mathcal{R}(x, y) \qquad cl_\mathcal{R}(cl_\mathcal{R}(S)) = \bigcup \{\mathcal{R}(x', y') \mid x', y' \in \bigcup_{x,y \in S} \mathcal{R}(x, y)\}$$

(a)   (b)   (c)

Fig. 4. Schematic comparison of (a) the convex hull closure, shaded in 'grey'; (b) recombination P-structure closure $cl_\mathcal{R}$, in 'solid lines'; and, (c) the twice-iterated $cl_\mathcal{R}^2$, in 'solid and dotted lines'. For a given subset $S := \{A, B, C, D\}$ of some space X.

Besides, recombination P-structures can be used to define closures that are equivalent to convex hulls [11,12]. One may attempt to define one by taking the union over all pairs of points for a given set, and collect their intervals given by a P-structure (Fig. 4b). Precisely, this is the closure $cl_\mathcal{R}(S) := \bigcup_{x,y \in S} \mathcal{R}(x, y)$, where $S \subseteq X$ for a general recombination P-structure $(X, \mathcal{R})$ [12]. However, it

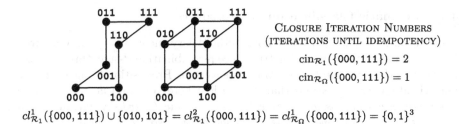

CLOSURE ITERATION NUMBERS
(ITERATIONS UNTIL IDEMPOTENCY)

$$\text{cin}_{\mathcal{R}_1}(\{000, 111\}) = 2$$

$$\text{cin}_{\mathcal{R}_\Omega}(\{000, 111\}) = 1$$

$$cl^1_{\mathcal{R}_1}(\{000, 111\}) \cup \{010, 101\} = cl^2_{\mathcal{R}_1}(\{000, 111\}) = cl^1_{\mathcal{R}_\Omega}(\{000, 111\}) = \{0, 1\}^3$$

Fig. 5. Space covered on the hypercube, with set of parents $\{000, 111\}$, by: one-point recombination closure with one iteration $cl^1_{\mathcal{R}_1}$ (left), uniform recombination closure $cl^1_{\mathcal{R}_\Omega}$ and one-point twice-iterated $cl^2_{\mathcal{R}_1}$ (center). Super-indices indicate number of closure iterations, and sub-indices the recombination P-structure chosen.

does not equal the convex hull (Fig. 4a) because, unlike convex hulls, the closure $cl_{\mathcal{R}}$ is not idempotent [7,15]. In general $cl_{\mathcal{R}}(cl_{\mathcal{R}}(S)) \neq cl_{\mathcal{R}}(S)$, compare Fig. 4b and c. See Fig. 5 for an example on the one-point recombination P-structure $\mathcal{R}_1$. Since $cl_{\mathcal{R}}$ is not idempotent, [12] suggests to define a recursive version, for some natural number $k \in \mathbb{N}_0$, as follows:

$$cl^k_{\mathcal{R}}(S) := \begin{cases} S & \text{if } k = 0, \\ cl_{\mathcal{R}}(S) = \bigcup_{x,y \in S} \mathcal{R}(x,y) & \text{if } k = 1, \\ cl^{k-1}_{\mathcal{R}}(cl_{\mathcal{R}}(S)) & \text{if } k \geq 2 . \end{cases} \tag{1}$$

The smallest k that makes $cl^k_{\mathcal{R}}$ idempotent is the *closure iteration number*[2], denoted $\text{cin}(S)$, of the subset $S \subseteq X$: $\text{cin}(S) := \min_{k \in \mathbb{N}_0} cl^{k+1}_{\mathcal{R}}(S) = cl^k_{\mathcal{R}}(S)$; and the *maximum closure iteration number* of X is $\text{cin}^*(X) := \max_{S \subseteq X} \text{cin}(S)$ [17].

However, what [12] does not explicitly notice is that $cl^k_{\mathcal{R}}$ (Eq. 1) is an equivalent definition of the convex hull closure, not just for recombination P-structures but more generally for 'relaxed' recombination P-structures satisfying only the null-recombination and symmetry axioms (also known as interval spaces).

Proposition 4 (Convex hull $\equiv$ recomb. P-structure closure [7,17]). Let $(X, \mathcal{R})$ be a recombination P-structure. Then, for any subset $S \subseteq X$ and any natural number $k \geq \text{cin}(S)$: $co(S) = cl^k_{\mathcal{R}}(S)$.

4 Main Results

This section contains the two main contributions of this paper. Section 4.1 shows that the geometric versus non-geometric crossovers classification can be expanded by incorporating new recombination P-structures classes. Section 4.2 formalises abstract convex search for recombination P-structure EAs.

---

[2] Certain crossover families (e.g. one-point and masked crossovers) have logarithmic upper-bounds for the closure iteration number, with increasing dimension of the search space [16]. In short, idempotency does not require excessively many iterations.

4.1   Axiomatic Classification of Crossovers

Proposition 1 justifies that GF is not a futile theory encompassing all crossovers, because not all of them are geometric. Since recombination P-structures may be a larger class than complete geometric crossovers, as Proposition 3 suggests, three non-trivial relevant questions that we consider here are: (Q1) *do recombination P-structures encompass all crossovers, thus making GF-ELs unification futile*; (Q2) *is there any crossover that is a recombination P-structure but not geometric*; and, (Q3) *how recombination P-structures and geometric crossover classes compare to each other.* We tackle them in order as follows: Lemmas 1 and 2 answer question (Q1) negatively; Theorem 1 together with Lemma 5 and Theorem 2 answer question (Q2) positively; and, Theorem 3 for question (Q3). The proofs of Lemmas 3, 5 and 6, and the proofs of Theorems 1 and 2 have been omitted due to page limit restrictions. The full version including proofs is available online.

Let us begin (Q1) by proving that Koza subtree swap crossover, which was proved to be non-geometric [8], is not a recombination P-structure.

Lemma 1. *Koza subtree swap crossover is not a recombination P-structure.*

Proof. It is not geometric since it does not fulfill the purity inbreeding property (i.e. recombining a parent with itself is the parent itself) [8]. Therefore, it does not fulfil the fix-point axiom of recombination P-structures $\mathcal{R}(x,x) = \{x\}$.   □

Davis order crossover is non-geometric [8]. Now, we prove it is not a recombination P-structure, because it fails the symmetry axiom (Example 5); and, neither for a symmetric version (Definition 9), because it is not size-monotonic (Example 6).

Definition 8 (Davis order crossover [18]). Let individuals $x, y, z \in S_n$ be permutations of length n of a finite set S. Pick at random loci i and j such that $1 \leq i \leq j \leq n$, as indices delimiting the crossover section[3]. Then, from the first parent x (the 'cutter') copy into offspring positions $z_i, \ldots, z_j$ the crossover section $x_i, \ldots, x_j$. Finally, from left to right, fill in order the remaining positions of z with the genes of the second parent y (the 'filler') except those that have been copied. We denote this crossover, parametrised with crossover section indices i and j, by $\text{DAVIS}(i,j) : S_n \times S_n \to S_n, (x,y) \mapsto z = \text{DAVIS}(i,j)(x,y)$.

Example 5 (Davis order crossover is asymmetric). Let parents $x :=$ 312 (the 'cutter') and $y := 123$ (the 'filler'). Consider offspring 2$\underline{1}$3 $=$ DAVIS$(2,2)(3\underline{1}2, 123)$, where $\underline{1}$ is the crossover section. Notice that 213 cannot be generated by DAVIS$(i,j)(123, 312)$ for any crossover section indices i and j, because it is not possible to generate 2 at the first position. Either the second parent places 3 or 1 as fillers, or the first parent places 1 with the crossover section. Therefore, DAVIS$(i,j)(312, 123) \neq$ DAVIS$(i,j)(123, 312)$, for $1 \leq i \leq j \leq 3$.

---

[3] Notice that a crossover section with all parent's genes is also a valid contiguous section.

Example 4 demonstrated that if a crossover is asymmetric, it does not mean there are no (symmetric) recombination P-structures based on it. In fact, that example shows how symmetry is enforced on recombination P-structures for traditional genetic crossovers (e.g. two-point or uniform crossovers) [5, 12]. Likewise, we may define a symmetric Davis order crossover.

Definition 9 (Symmetric Davis order crossover). Let arbitrary parents $x, y \in S_n$ be permutations of length n of a finite set S. Then, s-$\text{DAVIS}(x, y) := \bigcup_{1 \leq i \leq j \leq n} \{\text{DAVIS}(i, j)(x, y), \text{DAVIS}(i, j)(y, x)\}$ is what we call *symmetric Davis order crossover*; producing all possible offspring of parents x and y, over all possible crossover sections.

Example 6 (Symmetric Davis order crossover is not size-monotonic). Using the same parents as in the previous example, we have s-$\text{DAVIS}(312, 123)$ $= \{312, 123, 321, 132, 213\}$. That is $|\text{s-}\text{DAVIS}(312, 123)| = 5$. Recombining now parent $x := 312$ with offspring $z := 213$ we have s-$\text{DAVIS}(312, 213) = \{231\} \cup$ s-$\text{DAVIS}(312, 123)$. Therefore, $|\text{s-}\text{DAVIS}(312, 213)| = 6 > 5 = |\text{s-}\text{DAVIS}(312, 123)|$, so s-$\text{DAVIS}$ is not size-monotonic.

Lemma 2. Neither Davis order crossover (DAVIS) nor its symmetric version (s-DAVIS) are recombination P-structures.

Proof. Examples 5 and 6 provide counterexamples respectively.   □

Therefore, the answer to (Q1) is negative by Lemmas 1 and 2, because GF-EL unification is not a theory encompassing all crossovers. Regarding question (Q2), we show next two examples of crossovers that are not geometric but are recombination P-structures, at least in Hamming graphs: the all-paths crossover (Definition 10), and the intersecting-balls crossover (Definition 11). Note that these crossovers are not restricted to a particular representation, like UNIFORM crossover on strings, so one must be specified when proving whether they are geometric crossovers or recombination P-structures: choose a graph for all-paths, or a metric space for intersecting-balls.

Definition 10 (All-paths crossover [11]). Let G be a connected graph with vertex set $V(G)$. Let also $x, y, z \in V(G)$. Then, we define $\text{ALLPATHS}(x, y) := \{z \mid z \text{ lies on any } x\text{-}y \text{ path in } G\}$, as the *all-paths crossover* that returns offspring z lying on any path between parents x and y.

Example 7 (All-paths crossover in hypercubes). Let $(\mathcal{H}_2^3, d_{\text{H}})$ be the Hamming metric space of the three-dimensional hypercube graph. If parents are identical, offspring equal parents: $\text{ALLPATHS}(001, 001) = 001$. But, if the parents are different, then offspring are the whole vertex set: $\text{ALLPATHS}(000, 001) = \{0, 1\}^3$.

Lemma 3. Let $x, y \in \mathcal{H}_q^n$ be distinct vertices of a Hamming graph. Then, $\text{ALLPATHS}(x, y)$ returns all sequences $\mathcal{H}_q^n$ as offspring.

Proof. Omitted due to space restrictions. See the full paper version for details.

Theorem 1. Let $(\mathcal{H}_q^n, d_\mathrm{H})$ be the Hamming metric space of an arbitrary Hamming graph. Then, $\mathrm{ALLPATHS}(x,y)$ is not a geometric crossover, but it is a recombination P-structure, for any parents $x, y \in \mathcal{H}_q^n$.

Proof. Omitted due to space restrictions. See the full paper version for details.

Definition 11 (Intersecting-balls crossover). Let (X,d) be any metric space and arbitrary parents $x, y \in X$. Then, we define the *intersecting-balls crossover* as $\mathrm{BALLS}(d) : X \times X \to S \subseteq X$, $(x,y) \mapsto B_d[x; d(x,y)] \cap B_d[y; d(y,x)]$, parametrised with metric d.

Lemma 4 is necessary to justify that [19] used an equivalent definition of the intersecting-balls crossover, which they suggested to be geometric when in fact it is not necessarily geometric (Lemma 5).

Lemma 4. Rothlauf's crossover $R(x,y) := \{z \mid \max\{d(x,z), d(z,y)\} \leq d(x,y)\}$ is identical to $\mathrm{BALLS}(d)(x,y)$, for any metric space (X,d) and any $x, y, z \in X$.

Proof. Follows directly: $z \in R(x,y) \iff d(x,z) \leq d(x,y)$ and $d(z,y) \leq d(x,y)$ $\iff z \in (B_d[x; d(x,y)] \cap B_d[y; d(y,x)]) \iff z \in \mathrm{BALLS}(d)(x,y)$. □

Definition 12 (Hamming intersecting-balls crossover). Let $(\mathcal{H}_q^n, d_\mathrm{H})$ be any Hamming metric space and any $x, y \in \mathcal{H}_q^n$. Then, the *Hamming intersecting-balls crossover* is $\mathrm{BALLS}(d_\mathrm{H})(x,y) := B_{d_\mathrm{H}}[x; d_\mathrm{H}(x,y)] \cap B_{d_\mathrm{H}}[y; d_\mathrm{H}(y,x)]$.

Lemma 5. Let $(\mathcal{H}_q^n, d_\mathrm{H})$ be any Hamming metric space. Then, $\mathrm{BALLS}(d_\mathrm{H})$ is not a geometric crossover.

Proof. Omitted due to space restrictions. See the full paper version for details.

Lemma 6. Let (X,d) be any metric space. Then, for all $x, y \in X$ it holds: if $z \in [x; y]_d$, then $\mathrm{BALLS}(d)(x,z) \subseteq \mathrm{BALLS}(d)(x,y)$.

Proof. Omitted due to space restrictions. See the full paper version for details.

Theorem 2. Let $(\mathcal{H}_q^n, d_\mathrm{H})$ be the Hamming metric space. Then, $(\mathcal{H}_q^n, \mathrm{BALLS}(d_\mathrm{H}))$ is a recombination P-structure.

Proof. Omitted due to space restrictions. See the full paper version for details.

Using the previous results we now compare recombination P-structures with geometric crossover classes (Q3).

Theorem 3 (Crossover classes). Let $\mathcal{RP}$ be the set of recombination P-structures (resp. $\overline{\mathcal{RP}}$ for crossovers that are not recombination P-structures), $\mathcal{GX}$ be the geometric crossovers, and $\mathcal{GX}$-complete be the complete geometric crossovers. It holds that: (a) $\overline{\mathcal{RP}} \neq \emptyset$; (b) $\mathcal{GX} \neq \mathcal{RP}$; (c) $\mathcal{RP} \cap \overline{\mathcal{GX}} \neq \emptyset$; (d) $\mathcal{GX}$-complete $\subset \mathcal{RP}$; and (e) $\overline{\mathcal{RP}} \subset \overline{\mathcal{GX}}$-complete.

Proof.(a) Lemmas 1 and 2 prove, respectively, that neither Koza subtree swap
nor (symmetric) Davis order crossovers are recombination P-structures.

(b) Example 2 shows that the single-offspring one-point crossover (ONEPOINT)
is geometric, but it is not a recombination P-structure due to its asymmetry.

(c) In Hamming graphs, Theorem 1 proves it for the ALLPATHS crossover, also
Lemma 5 and Theorem 2 prove it for BALLS(d_H) crossover.

(d) Follows from Proposition 3, and (c) makes the inclusion strict: complete
geometric crossovers are a subclass of geometric crossovers (Proposition 1).

(e) Let $\mathcal{U}$ be the universal set of crossovers, where $\mathcal{GX}$-complete, $\mathcal{RP} \subset \mathcal{U}$.
Then, using (d) and taking complements, $\overline{\mathcal{RP}} \subset \overline{\mathcal{GX}}$-complete follows.

$\square$

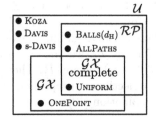

Fig. 6. Classification within the universal set $\mathcal{U}$ of crossovers by geometric crossovers
$\mathcal{GX}$ and recombination P-structures $\mathcal{RP}$, with class members examples: all-paths
(ALLPATHS), Hamming intersecting-balls (BALLS(d_H)), Davis order (DAVIS), symmetric Davis order (s-DAVIS), Koza subtree swap (KOZA), single-offspring one-point
(ONEPOINT) and uniform (UNIFORM) crossovers.

The main purpose of Theorem 3, summarised in Fig. 6, is to start developing
a unified and systematic understanding of crossovers within the geometric GF
(Sect. 2) and algebraic EL (Sect. 3) theories. By relaxing the axioms of recombination P-structures or properties of geometric crossovers, new formal crossover
subclasses can be incorporated, which can give us deeper insight not only into the
similarities and differences in behaviour of crossovers, but also about the theories themselves. For instance, EL landscape analysis is possible when a crossover
belongs to $\mathcal{GX}$-complete or $\mathcal{RP}$, but not necessarily for all those in $\overline{\mathcal{GX}}$-complete
(e.g. ONEPOINT) [1]. It should be clear that the focus of this classification is not
on the practical value of crossovers (e.g. KOZA is useful, however $\mathcal{GX}$ and $\mathcal{RP}$
miss it), but what crossovers can and cannot be studied within GF and ELs.

4.2   Algebraic Abstract Convex Search

We switch our attention from recombination P-structures themselves (Sect. 4.1)
to characterising the population-behaviour of EAs using them. Specifically, we
prove that any recombination P-structure EA, using any selection and replacement strategies but no mutation, exhibits an abstract convex search behaviour
analogue to Proposition 2.

Theorem 4 (Algebraic abstract convex search). Let $(X, \mathcal{R})$ be a recombination P-structure and $P_t \subseteq X$ the population at generation t. Then, there exists a maximum closure iteration number $m := \operatorname{cin}^*(X)$ such that: $cl_{\mathcal{R}}^m(P_0) \supseteq cl_{\mathcal{R}}^m(P_1) \supseteq \cdots \supseteq cl_{\mathcal{R}}^m(P_t) \supseteq cl_{\mathcal{R}}^m(P_{t+1})$; for any EA repeating the cycle: selection, recombination $\mathcal{R}$ and replacement.

Proof. We need to prove that: (a) such maximum closure iteration number exists; and, (b) the closure of recombination P-structures produces a nested chain of inclusions of population sets.

(a) From Proposition 4 we know that for any population $P_t \subseteq X$ there exists a natural number $k \geq \operatorname{cin}(P_t)$, such that $co(P_t) = cl_{\mathcal{R}}^k(P_t)$. Since all populations P_t are subsets of X, the maximum closure iteration number $m := \operatorname{cin}^*(X)$ exists. By definition (see below Eq. 1), $\operatorname{cin}^*(X) \geq \operatorname{cin}(P_t)$ for all populations P_t. Therefore, $co(P_t) = cl_{\mathcal{R}}^m(P_t)$ for all populations P_t.

(b) Since (a) proves that there exists a maximum closure iteration number such that $co(P_t) = cl_{\mathcal{R}}^m(P_t)$ for any P_t, we know also that $co(P_t) \supseteq co(P_{t+1})$ if and only if $cl_{\mathcal{R}}^m(P_t) \supseteq cl_{\mathcal{R}}^m(P_{t+1})$ for any P_t. Therefore, a chain of inclusions is formed, provided that selection and replacement can only return a subset of the convex hull of a population [10]. $\qquad\square$

Example 8 (Abstract convex search of all-paths crossover). ALLPATHS convex search is 'degenerate' because, as long as the current population contains different parents, the vertex set of the Hamming graph is returned as offspring; until other mechanism like replacement or selection returns a population with the same individual. For instance, if $P_1 := \{001, 010, 111\}$, the following chain may be generated: $cl_{\text{ALLPATHS}}^1(P_1) = \{0, 1\}^3 \supseteq \{0, 1\}^3 \supseteq \ldots \supseteq \{0, 1\}^3 \supseteq 001$. Notice that the maximum closure iteration number of ALLPATHS is one, because ALLPATHS generates at once the whole vertex set or a singleton set of the Hamming graph, which are convex sets [17].

It remains as future work to find whether 'degenerate' cases like Example 8 are the <u>only</u> kind of algebraic abstract convex search for crossovers in the class $\mathcal{RP} \cap \overline{\mathcal{GX}}$ (i.e. non-geometric crossovers that are recombination P-structures).

Note that Proposition 2 and Theorem 4 become equivalent for complete geometric crossovers in finite metric spaces, since they are a subclass of recombination P-structures (Theorem 3). Table 1 illustrates it for the uniform recombination P-structure $\mathcal{R}_\Omega$, which is a complete geometric crossover, whose maximum closure iteration number is one for all populations P: $co(P) = cl_{\mathcal{R}_\Omega}^1(P)$ [11].

Theorem 4 has significant implications for GF and EL. First, it shows that the convex search of complete geometric-crossover EAs can be phrased algebraically in EL via recombination P-structures, in a way that remains consistent with GF. Secondly, that the abstract convex behaviour of some non-geometric crossovers, for example the ALLPATHS crossover (Definition 10), can be formally justified. Finally, we now have a proper characterisation of population behaviour for recombination P-structures EAs, which is less restrictive than the previous crossover-walk model (Sect. 3.1) because it is defined for general recombination

Table 1. A small trace of abstract convex search for the three-dimensional binary Hamming space using uniform recombination with initial population $\{001, 010, 111\}$, population size equal three, generational replacement and no selection.

Parents population	Parents closure (and their schema)	Offspring population	Hyperedge of parents closure
$P_1 = \{001, \ 010, \ 111\}$	$cl^1_{\mathcal{R}_\Omega}(P_1) = \{0,1\}^3$ (***)	$\{010, \ 100, \ 110\}$	
$P_2 = \{010, \ 100, \ 110\}$	$cl^1_{\mathcal{R}_\Omega}(P_2) = \{0,1\}^2 \times \{0\}$ (**0)	$\{100, \ 000, \ 100\}$	
$P_3 = \{100, \ 000, \ 100\}$	$cl^1_{\mathcal{R}_\Omega}(P_3) = \{000, 100\}$ (*00)	$\{100, \ 100, \ 000\}$	
$P_4 = \{100, \ 100, \ 000\}$	$cl^1_{\mathcal{R}_\Omega}(P_4) = \{000, 100\}$ (*00)	$\{100, \ 100, \ 100\}$	
$P_5 = \{100, \ 100, \ 100\}$	$cl^1_{\mathcal{R}_\Omega}(P_5) = 100$		

P-structures not just string crossovers, and no assumptions are made on selection pressure or genes interdependence (epistasis). Moreover, it has a clear geometrical interpretation (Sect. 2.2) and GF has demonstrated that is analytically useful to study recombination-based EAs [2]. As a final remark, some of our results (e.g. Proposition 3 or Theorem 4) can be further generalised to more abstract algebraic spaces like interval spaces [7], by relaxing the axioms of recombination P-structures and keeping only the nullary and symmetric ones (Definition 6). It turns out that Proposition 4, which is key for Theorem 4, remains true [7]. However, our aims here have not required such degree of generality.

5   Summary and Future Work

Previous work [1] established the foundations for developing a unified approach to analyse fitness landscapes, based on an algebraic theory of fitness landscapes (Sect. 3) and a geometric framework of EAs (Sect. 2). Aiming for a solution to an open challenge in evolutionary computing: to determine a priori for a given problem if its fitness landscape matches a class of EAs where good performance can be guaranteed across representations. However, to use EL as an analytic tool, we needed to show that recombination landscapes seen from EL and GF can be unified, that is their underlying space structures are not essentially different.

This paper expands the unification between GF and EL to eventually develop an integrated GF-EL framework; where EAs can be analysed and designed across representations (GF), but also it is possible to know analytically if good performance can be guaranteed for a given problem landscape (EL). First, we augmented the classification of crossovers in GF with algebraic crossover classes from EL (Sect. 4.1). Then, we formalised the abstract convex population behaviour of recombination P-structure EAs, as an extension of complete geometric-crossover EAs, which provides a more general characterisation of population behaviour than the existing model in EL (Sect. 4.2).

As future work, we consider the following questions. First, to augment the classification of crossovers with new classes or members and compare them. Secondly, to study the relation between abstract convex spaces (e.g. convex geometries [9]) and crossovers, by looking at how the presence or absence of convexity influences crossover behaviour. Thirdly, to export the runtime analysis of geometric-crossover EAs [2] to recombination P-structure EAs; and, conversely, use the ideas presented here to derive new insights about runtime analysis of geometric-crossover EAs. Finally, it would be interesting to integrate mutation within GF-EL unification to understand more realistic EAs using mutation and recombination operators.

References

1. García, M.D., Moraglio, A.: Bridging elementary landscapes and a geometric theory of evolutionary algorithms: first steps. In: Auger, A., Fonseca, C.M., Lourenço, N., Machado, P., Paquete, L., Whitley, D. (eds.) PPSN 2018, Part II. LNCS, vol. 11102, pp. 194–206. Springer, Cham (2018). https://doi.org/10.1007/978-3-319-99259-4_16
2. Moraglio, A., Sudholt, D.: Principled design and runtime analysis of abstract convex evolutionary search. Evol. Comput. **25**(2), 205–236 (2017)
3. Moraglio, A.: Towards a Geometric Unification of Evolutionary Algorithms. Doctoral thesis, University of Essex, Essex, UK, November 2007
4. Stadler, P.F.: Towards a theory of landscapes. In: López-Peña, R., Waelbroeck, H., Capovilla, R., García-Pelayo, R., Zertuche, F. (eds.) Complex Systems and Binary Networks. Lecture Notes in Physics, vol. 461, pp. 78–163. Springer, Heidelberg (1995). https://doi.org/10.1007/BFb0103571
5. Stadler, P.F., Wagner, G.P.: Algebraic theory of recombination spaces. Evol. Comput. **5**(3), 241–275 (1998)
6. Stadler, P.F., Seitz, R., Wagner, G.P.: Population dependent fourier decomposition of fitness landscapes over recombination spaces: evolvability of complex characters. Bull. Math. Biol. **62**(3), 399–428 (2000). Society for Mathematical Biology
7. van de Vel, M.L.J.: Theory of Convex Structures. North-Holland Mathematical Library, North-Holland (1993)
8. Moraglio, A., Poli, R.: Inbreeding properties of geometric crossover and non-geometric recombinations. In: Stephens, C.R., Toussaint, M., Whitley, D., Stadler, P.F. (eds.) FOGA 2007. LNCS, vol. 4436, pp. 1–14. Springer, Heidelberg (2007). https://doi.org/10.1007/978-3-540-73482-6_1

9. Farber, M., Jamison, R.E.: Convexity in graphs and hypergraphs. SIAM J. Algebr. Discret. Meth. **66**(3), 231–247 (1986). Society for Industrial and Applied Mathematics

10. Moraglio, A.: Abstract convex evolutionary search. In: Proceedings of the 11th Workshop on Foundations of Genetic Algorithms, FOGA 2011, pp. 151–162. ACM, Schwarzenberg, Austria (2011)

11. Changat, M., et al.: Topological Representation of the Transit Sets of k-Point Crossover Operators. arXiv:1712.09022 (2017)

12. Gitchoff, P., Wagner, G.P.: Recombination induced hypergraphs: a new approach to mutation-recombination isomorphism. Complexity **2**(1), 37–43 (1996)

13. Moraglio, A., Togelius, J., Silva, S.: Geometric differential evolution for combinatorial and programs spaces. Evol. Comput. **21**(4), 591–624 (2013)

14. Holland, J.H.: Adaptation in Natural and Artificial Systems: An Introductory Analysis With Applications to Biology, Control, and Artificial Intelligence. Complex Adaptive Systems. MIT Press, Cambridge (1992, reprint edn)

15. Stadler, B.M.R., Stadler, P.F.: Generalized topological spaces in evolutionary theory and combinatorial chemistry. J. Chem. Inf. Comput. Sci. **42**(3), 577–585 (2002). American Chemical Society

16. Mitavskiy, B.: Crossover invariant subsets of the search space for evolutionary algorithms. Evol. Comput. **12**(1), 19–46 (2004)

17. Pelayo, I.M.: Geodesic Convexity in Graphs. Springer Briefs in Mathematics. Springer, New York (2013). https://doi.org/10.1007/978-1-4614-8699-2

18. Davis, L.: Applying adaptive algorithms to epistatic domains. In: Proceedings of the 9th International Joint Conference on Artificial Intelligence-Volume 1, IJCAI 1985, pp. 151–162. Morgan Kaufmann Publishers Inc. (1985)

19. Thorhauer, A., Rothlauf, F.: On the bias of syntactic geometric recombination in genetic programming and grammatical evolution. In: Proceedings of the 2015 Annual Conference on Genetic and Evolutionary Computation, GECCO 2015, pp. 1103–1110. ACM, New York, NY, USA (2015)

Program Trace Optimization
with Constructive Heuristics
for Combinatorial Problems

James McDermott[1]([⊠]) and Alberto Moraglio[2]

[1] National University of Ireland, Galway, Ireland
james.mcdermott@nuigalway.ie
[2] University of Exeter, Exeter, UK
A.Moraglio@exeter.ac.uk

Abstract. Program Trace Optimisation (PTO), a highly general optimisation framework, is applied to a range of combinatorial optimisation (COP) problems. It effectively combines "smart" problem-specific constructive heuristics and problem-agnostic metaheuristic search, automatically and implicitly designing problem-appropriate search operators. A weakness is identified in PTO's operators when applied in conjunction with smart heuristics on COP problems, and an improved method is introduced to address this. To facilitate the comparison of this new method with the original, across problems, a common format for PTO heuristics (known as generators) is demonstrated, mimicking GRASP. This also facilitates comparison of the degree of greediness (the GRASP α parameter) in the heuristics. Experiments across problems show that the novel operators consistently outperform the original without any loss of generality or cost in CPU time; hill-climbing is a sufficient metaheuristic; and intermediate levels of greediness are usually best.

Keywords: Constructive heuristics · GRASP · Search operators

1 Introduction

Many heuristic and metaheuristic methods have been applied in the field of combinatorial optimisation (COP). Often they achieve good results on problems where exact methods become infeasible. Research in this area is somewhat disunified in that many methods have been individually specialised to many problems: in each case, a combination of domain expertise and algorithmic expertise is required to design suitable representations and operators.

However, many of these methods do share common concepts, such as (in constructive heuristics) sampling of solution elements biased by their cost, and (in metaheuristics) perturbation of existing solutions. It is a natural goal to unify these methods in a single, general approach. It is also desirable to achieve a degree of automation in the application of these methods to problems, rather than requiring algorithmic expertise to be re-applied to each new problem.

© Springer Nature Switzerland AG 2019
A. Liefooghe and L. Paquete (Eds.): EvoCOP 2019, LNCS 11452, pp. 196–212, 2019.
https://doi.org/10.1007/978-3-030-16711-0_13

Program trace optimisation (PTO) [1] is a recent optimisation framework which is highly general and unifying, and which does (in a sense to be clarified later) automate the work of adapting an algorithm to a problem; and so it responds to these research challenges.

In fact, PTO is not just highly general but in a sense *maximally* general, because it uses a representation which by the Church-Turing thesis is the most general possible – the *program trace*, that is a data structure representing a history of execution of a program. The program in question is a non-deterministic *solution generator*, which plays the role of a genotype-phenotype map, seeing the program trace as a genotype and a candidate solution as a phenotype. To be specific, the generator is a program or function which randomly samples one element from the solution space, with or without bias, for example in a bitstring space it returns a single bitstring chosen randomly. With bias, it becomes a constructive or generative heuristic. PTO thus gives a unifying view.

Heuristic and metaheuristic approaches are commonly hybridised, for example in GRASP [2]. In PTO, both the heuristic and metaheuristic aspects are pluggable, and it is natural in PTO to compare a given heuristic across several metaheuristics. In this paper, we borrow well-known constructive heuristics for several problems as PTO generators. This re-use of known-good ideas is a central goal of PTO. We write all of them in a GRASP-like format, and this gives the ability to understand and compare PTO program traces and search behaviour across problems. It allows for a systematic experiment also on the well-understood "greediness" parameter of GRASP.

In most metaheuristics, the ability to make *small* perturbations is essential to success: the perturbation operator should give a new solution which retains or *re-uses* almost all of the information of the previous solution. But on some COP problems, with typical constructive heuristics as generators, it turns out that the PTO operators do not succeed in re-using as much of this information as they could. The problem is due to the repair mechanism which runs after each genetic variation operator. Therefore, in this paper an improved repair mechanism is designed, which succeeds in re-using as much information as possible.

Reader's Guide. In the next section, Sect. 2, we describe PTO itself and previous related work, with an expanded explanation of PTO mechanisms. We then describe in detail our fixed format for generators for COP problems in Sect. 3. The novel repair mechanism is described in Sect. 4. Experiments and results are in Sect. 5, and we conclude with Sect. 6.

2   PTO and Related Work

PTO uses a user-supplied non-deterministic generator function as an implicit definition of the representation. PTO sees the sequence of random decisions made during execution of this program as a genotype.

PTO's modular design is shown in Fig. 1. The user supplies a problem-specific generator and fitness function. The core of PTO executes the generator once

Fig. 1. PTO architecture

per individual. Each execution gives both a trace (genotype) and a solution (phenotype). Any metaheuristic can be plugged-in as the solver, carrying out its genetic operations on the trace.

The idea of using such a sequence of decisions as a genotype was introduced in the Program Optimisation with Dependency Injection (PODI) system [3]. The generators used varied from uniform sampling of the space for symbolic regression problems, to initialisation operators from previous metaheuristic research in communications network design, re-purposed as generators, to idiosyncratic hand-written code originally intended only as a tool for random generation of 3D structures. PTO goes beyond PODI by using a more sophisticated trace representation, allowing improved re-use of genetic material.

The idea of hybridising constructive heuristics with metaheuristic search is well-established in the COP literature. GRASP combines a greedy heuristic with path relinking and local search [2]. As a generalisation, biased randomisation and simulated annealing have been combined [4,5]. PTO with GRASP-like generators differs from GRASP itself by allowing any metaheuristic to be plugged-in. In some hybrid approaches, the greedy constructive procedure is used as a "smart" initialisation, after which the algorithm uses metaheuristic search alone [6]. PTO uses the heuristic throughout the metaheuristic search process.

In the evolutionary computation literature, the genotype-phenotype mapping is a common research topic. "Smart" mappings seek to build problem-awareness (e.g. constraint handling) directly into the algorithm, e.g. [7]. PTO differs from such approaches by using a program trace as the genotype and aims more for universality and modularity – the algorithm is unchanged for new problems.

2.1 The Program Trace: A Universal Solution Representation

As already explained, in PTO the program trace is the sequence of outcomes of (random) decisions made by the generator in producing a particular solution. The trace can be manipulated: it can be "played back" in the generator to redo the same sequence of decisions and produce the same solution; it can be edited and played back to produce a variant solution; two parent traces can be combined and the result played back to produce a child solution. That is, the trace is a genotype, the solution is the corresponding phenotype, and the playback mechanism in the generator is a developmental mapping; editing and recombination of traces are search operators.

The trace is a "universal representation" that applies to any problem because it is implicit in the problem definition (in the generator) and can be extracted automatically by tracing. No other representation can be more general or more powerful, since the generator can use Turing-complete code. Metaheuristics defined on such a representation can be applied *unchanged* to any problem, thus becoming universal optimisers.

The trace representation is a *dictionary*. Each entry is a key-value pair corresponding to one random decision made by a random function call during execution of the generator. The value is simply the output of the random function call, for example an integer or float. The key gives the structural position of that decision in the execution trace, rather like a stack address: it includes the list of functions, their line numbers, and their loop indices, that precede the random call in the call stack. This scheme follows the approach of Wingate et al. [8]. Examples in the context of well-known COP problems are given in Sect. 3.

2.2 Solvers and Search Operators

The search operators required by metaheuristics such as hill-climbers and evolutionary algorithms are defined on the trace representation. They are defined in a principled way based on the geometric framework [9].

Initialisation runs the generator and traces its execution. The resulting trace becomes the newly-initialised genotype, and the output of the generator becomes its phenotype. **Point mutation** picks a random entry of the trace and replaces its value with a value drawn from the same random call. **Uniform crossover** aligns parent traces on their names (i.e. dictionary keys). For names that appear in both parents, the offspring inherits the corresponding entries from either parent at random, i.e. using a random mask to select. For names that appear in only one parent, the offspring inherits all of them.

Repair is applied after each alteration of the trace, i.e., after the application of any variation operator. The repair takes place when running the modified trace in the generator in playback mode to generate the corresponding solution. If there is a mis-match, i.e. the current value comes from a random call other than the one identified by the name, then a new random value is drawn from the correct random call. If the trace is used up before the generator finishes, the trace is extended with new random entries as needed. Excess entries in the trace, not used by the generator, are deleted. Repair is discussed further in Sect. 4.

Given these operators, metaheuristic solvers such as random search, hill-climbing, and a GA can be defined. The trace becomes the genotype, and the generator becomes the genotype-phenotype map.

2.3 The Role of the Generator

The PTO generator can be seen as a genotype-phenotype mapping, where the genotype is the program trace and the phenotype is a candidate solution. PTO search operators work directly on the genotype, but we can also see them as implicitly working on the phenotype. What is their behaviour at that level?

In the framework of Jones [10], a search problem begins with an *object space* – in which the possible solutions are contained. The user is required to provide a representation space, that is a space in which points can be manipulated by search operators, and points can be mapped "forward" to the object space. E.g. in the TSP, the object space is the space of tours, and the representation space is the space of permutations. The distinction captures the extra meaning and structure associated with the object space. Objects in the object space are human-readable; those in the representation space, machine-manipulable. In PTO, the generator *implicitly* defines the object space. The job of the PTO user is to supply a generator which samples from the object space, rather than to design a representation space and operators on it. This workflow may suit domain experts better (but may be less familiar to metaheuristics researchers). One feature of PTO which is interesting for constrained COP problems such as JSSP is that it searches the space of *feasible* solutions. For many such problems it is not difficult to write a generator giving only feasible solutions, but it is difficult to design mutation and crossover operators that given feasible parents are guaranteed to return feasible offspring. PTO's search operators preserve feasibility automatically.

The operators applied by PTO on the trace representation correspond, when viewed at the level of the solution, to operators well-designed for the problem, in the sense that they take advantage of the meaning and structure of the object space. For example, if the solution space is of bitstrings, then a natural generator gives operators equivalent to well-known GA bitstring operators. If a generator uses nested loops, the implicit representation is a matrix; if one uses recursion the implied representation is a tree. Thus the same operators on traces result in meaningful operators for vectors, matrices and trees. This has been demonstrated in previous work [1], giving confidence that the PTO implicit adaptation works well. The case of permutations is different and is treated in Sect. 4. For a further example, when a greedy, randomised constructive heuristic is used as the generator, PTO can be expected to behave similarly to mutate-and-repair methods already well-known in the COP literature [4,5]. Although the end result may replicate a known-good method, the benefit for researchers and practitioners is in the automation: good problem-appropriate operators are derived, requiring domain knowledge but not algorithmic knowledge on the part of the user.

However, some caveats apply. The operators designed by PTO will be well-designed for the meaning and structure of the object space only to the extent that this meaning and structure is present in the user-supplied generator. Two generators which are semantically identical but syntactically different may induce different operators. As an example, given a list L from which the generator must sample an element, `i = random.choice(range(len(L))); x = L[i]` is semantically equivalent to `x = random.choice(L)`, but the latter expresses problem structure more directly. It is an assumption of PTO that the user will use direct formulations such as this. Also, in its implicit design of operators, PTO uses only the generator, not the objective function. For example, both the n-queens problem and the TSP can be represented as permutation problems, so a generator which uniformly samples permutations could be used equally for both. It would give no advantage on either problem relative to a standard permutation representation. A domain expert might see that for the TSP, a more suitable generator might iteratively build a solution heuristically guided by inter-city distance. This is a method of building problem knowledge into the representation.

2.4 Open Questions

Several important research questions remain open. (1) As described, PTO automatically designs operators for each problem. We wish to explore how well PTO performs across COP problems with the minimum input of expertise. (2) A beneficial side-effect of the PTO unifying view is the ability to compare across problems. To further this comparison, in the current paper we make use of a fixed format, borrowed from the core idea of GRASP, for the solution generators for several different problems. These generators are parameterised by a single parameter α which controls greediness. Furthermore, in PTO any metaheuristic can be used in combination with any generator, including combinations which may be rare in the COP literature. We wish to compare performance and the obtained traces and solutions, across problems, searchers and α values. (3) PTO relies for its success on the ability to manipulate the trace data structure. As we will see, in some COP problems the natural generator leads to a trace in which, after manipulation, many elements of the trace cannot be re-used, and so learning is lost. We introduce to PTO a novel mechanism to prevent this and allow strong re-use of trace elements, and we investigate its effect. We do not aim in this paper to compare PTO performance with other approaches.

3 GRASP in PTO

To apply PTO to solve COP problems, we need to provide a problem-specific solution generator. In this paper we use a fixed GRASP-like generator scheme for all problems, as shown in Algorithm 1. It assumes that problem solutions are characterised as aggregations of features: for example, in TSP the cities are features and a solution is a list of cities. At each step, it finds a list of allowed features (a subset of those not yet chosen); calculates their per-feature cost in the presence of the in-progress solution; filters them for low cost, forming the *restricted candidate list* (RCL); and chooses randomly.

Algorithm 1 is generic, suitable for several COP problems. It remains to fill in its sub-procedures `empty-solution()`, `complete()`, `allowed-features()`, `cost-feature()`, and `add-feature()` in a problem-specific way. These procedures are shown for several problems in Algorithms 2–4. For TSP, only the `cost-feature` procedure is shown as the others are the same as for Ordering. JSSP procedures are omitted due to lack of space but are specified in our source code available online.

The only source of randomness in our generators is add-feature() (line 9 of Algorithm 1). This is the only part of the generator affecting the trace representation, hence we will have a uniform, sequential trace structure for all problems. The parameter α controls the level of greediness in the heuristic ($\alpha = 0$ is fully greedy, $\alpha = 1$ is fully random), and has the same interpretation across problems (and in previous GRASP literature). As we will see, this fixed generator format also leads to a fixed trace format, aiding our understanding of the algorithm.

Algorithm 1. GRASP's greedy randomised construction scheme

```
 1: procedure greedy-randomised-construction()
 2:     solution ← empty-solution()
 3:     while not complete(solution) do
 4:         features ← allowed-features(solution)
 5:         for f in features do
 6:             costs[f] ← cost-feature(solution, f)
 7:         end for
 8:         RCL ← { f | costs[f] ≤ min(costs) + α(max(costs) - min(costs))}
 9:         solution ← add-feature(solution, random.choice(RCL))      ▷ Randomness
10:     end while
11:     return solution
12: end procedure
```

Algorithm 2. Procedures for the Ordering problem on n items

```
 1: procedure empty-solution()
 2:     return empty list
 3: end procedure
 4: procedure allowed-features(solution)
 5:     return {i | 1 ≤ i ≤ n ∧ i ∉ solution}
 6: end procedure
 7: procedure cost-feature(solution, f)
 8:     if solution is empty then
 9:         last-item ← 0
10:     else
11:         last-item ← last element of solution
12:     end if
13:     return |f−last-item+1|                    ▷ equal to 0 for consecutive integers
14: end procedure
15: procedure complete(solution)
16:     return solution has length n?                          ▷ True or False
17: end procedure
18: procedure add-feature(solution, f)
19:     return append f to solution
20: end procedure
```

Algorithm 3. New procedures for TSP on n cities with distance matrix D

```
1: procedure cost-feature(solution, f)
2:     if solution is empty then
3:         return 0                      ▷ all cities have zero cost as start city
4:     else
5:         last-item ← last element of solution
6:     end if
7:     return D[last-item, f]            ▷ travel cost from current to next city
8: end procedure
```

Algorithm 4. Procedures for a Knapsack problem with items $1 \ldots n$, weights w, profits p, and weight limit W

```
procedure allowed-features(solution)
    remaining ← {i | 1 ≤ i ≤ n ∧ i ∉ solution}
    current ← ∑{w[i] | i ∈ solution}
    return {i | i ∈ remaining ∧ current + w[i] ≤ W}
end procedure
procedure cost-feature(solution, f)
    return −p[f]                ▷ negative: p is a profit but we must return a cost
end procedure
procedure complete(solution)
    current ← ∑{w[i] | i ∈ solution}
    min-weight-left ← min{w[i] | i ∉ solution}
    return solution has length n? or current + min-weight-left > W
end procedure
```

3.1 Examples of Genotype-Phenotype Correspondence

In the following, we give two examples to illustrate the information contained in the trace representation (genotype), and how the same genotype corresponds to different types of solutions (phenotypes) for different problems.

Let us consider TSP with 10 cities and with $\alpha = 0.5$. The phenotype [0, 9, 2, 1, 6, 3, 4, 7, 5, 8] (tour of cities) corresponds to the trace shown in Table 1:

Table 1. Example trace for TSP on 10 cities with phenotype [0, 9, 2, 1, 6, 3, 4, 7, 5, 8].

Address	Type	Value
0	random.choice([0, 1, 2, 3, 4, 5, 6, 7, 8, 9])	0
1	random.choice([2, 4, 8, 9])	9
2	random.choice([1, 2, 3, 4, 5])	2
3	random.choice([1, 3, 4, 5])	1
4	random.choice([5, 6, 7])	6
5	random.choice([3, 4, 5])	3
6	random.choice([4, 5])	4
7	random.choice([7, 8])	7
8	random.choice([5])	5
9	random.choice([8])	8

The trace has three parts: (i) a runtime address (or name) of each entry in the trace; (ii) an entry type which consists of the elementary random generator and the values passed to it as argument; (iii) the value generated by the random generator. In all examples in this paper, the sequential nature of the GRASP generator means that the runtime address simplifies to an incrementing integer. The elementary random generator here is always `random.choice` as it is the only source of randomness in the GRASP generator. The argument passed to it is the RCL available at the moment of the call of the elementary random generator. This argument at different point in the execution is naturally different depending on the features still available at that point (e.g. cities not yet used in the construction of the solution) as well as features that have passed the selection based on the specific heuristic used in the construction (e.g. cities nearer to the last city of the tour under construction). The last column contains the actual values sampled by the elementary random generator with the specific argument. So, e.g., in the second line the value 9 was sampled from the call to `random.choice([2, 4, 8, 9])`.

As a second example, let us consider KNAPSACK with 10 items and with $\alpha = 1.0$ (fully random generator). This removes the effect of the heuristic, leaving only the constraining effects of previously selected items and backpack capacity.

The phenotype (knapsack items) `[9, 3, 1, 0, 2]` corresponds to the trace shown in Table 2:

Table 2. Example trace for KNAPSACK with 10 items with phenotype `[9, 3, 1, 0, 2]`.

Address	Type	Value
0	random.choice([0, 1, 2, 3, 4, 5, 6, 7, 8, 9])	9
1	random.choice([0, 1, 2, 3, 4, 5, 6, 7, 8])	3
2	random.choice([0, 1, 2, 4, 5])	1
3	random.choice([0, 2])	0
4	random.choice([2])	2

The interpretation of the trace is analogous to the TSP case. The TSP and Knapsack are different problems, but have the same trace representation as they use the same generator format, and in particular the same source of randomness in the same program execution context. The same is true of the other problems considered in this paper.

4   New Repair Method in PTO

In the following, we explain the current trace repair method in PTO, and why this needs refining, especially in conjunction with smart generators.

The trace in PTO is annotated with names and types on each entry. The name of an entry provides its 'execution address'. The type of an entry is the elementary generator and the arguments passed to it, which was used to generate the value in the trace at that entry.

Variation operators acting on the trace are typed, i.e., they change the values of the trace in conformity to their entry types, e.g., point mutation replaces the value at an entry in the trace with a new value obtained by re-sampling the associated elementary random generator to that entry (with the input arguments). However, this may still result in invalid traces, which present inconsistencies when expressed into phenotypes, because of the (implicit) 'runtime' dependencies between entries of the trace. For this reason, PTO has a repairing method that is applied to each trace when modified.

The current repairing method is as follows. When generating the corresponding phenotype of a trace (by playing back the trace in the generator), if we get a mismatch between the required type from the generator (i.e., the current elementary random function called in the generator) and the type of the corresponding entry in the trace (identified by its 'execution address'), then the entry of the trace is repaired by replacing its type with the required type, and discarding the value at that entry, which is then regenerated by sampling it from the new type (i.e., new random function and its argument) of the entry.

This repairing method correctly amends traces to produce valid phenotypes, but it may also be quite disruptive as it may require substantial change to a trace to produce a valid trace (see example below). This effectively would correspond to a form of macro-mutation (applied in addition to the intended modification done by variation operator) that may be detrimental for search performance. Ideally, we would like a repair method that produces a valid phenotype while making a minimal change to the trace. This is the aim of the new repair method.

Analogously to the old method, the new method repairs the type of an entry when a mismatch is encountered, however the value at that entry is kept (not re-sampled) if that value *could have been obtained* by running the elementary random function linked with the new entry type. For example, if the type and value of an entry in the trace is `random.choice([1,2,3])` : 2 and the required type from that generator is `random.choice([2,3,4])`, the value 2 at that entry can be "recycled" as it could have been obtained from `random.choice([2,3,4])`, and only the type of the entry is amended, giving `random.choice([2,3,4])` : 2.

The new repair method is much less disruptive in generators in which elementary random functions are called with arguments that depend on random outcomes in the generator at earlier stages. This is the situation that arises when using GRASP-in-PTO generators (but also in many other cases) in which the list of items available to be sampled at a point in time (the argument of the `random.choice` function) to be added to a solution under construction are those that have not been selected previously (as random outcomes at previous stages in the generator).

Using the old repair method, a single change in a value of the trace (point mutation) triggers a 'snowball' effect of type-mismatch in all the subsequent

Table 3. Point mutation (top) and 1-point crossover (bottom).

	Trace	Mutated trace	Old repair	New repair
1	[1,2,3,4,5,6,7] : 1	[1,2,3,4,5,6,7] : 1	[1,2,3,4,5,6,7] : 1	[1,2,3,4,5,6,7] : 1
2	[2,3,4,5,6,7] : 2	[2,3,4,5,6,7] : 7	[2,3,4,5,6,7] : 7	[2,3,4,5,6,7] : 7
3	[3,4,5,6,7] : 3	[3,4,5,6,7] : 3	[2,3,4,5,6] : ?	[2,3,4,5,6] : 3*
4	[4,5,6,7] : 4	[4,5,6,7] : 4	[2,3,4,5,6] : ?	[2,4,5,6] : 4*
5	[5,6,7] : 5	[5,6,7] : 5	[2,3,4,5,6] : ?	[2,5,6] : 5*
6	[6,7] : 6	[6,7] : 6	[2,3,4,5,6] : ?	[2,6] : 6*
7	[7] : 7	[7] : 7	[2,3,4,5,6] : ?	[2] : ?
	Phenotype		Phenotype	Phenotype
	(1 2 3 4 5 6 7)		(1 7 ? ? ? ? ?)	(1 7 3 4 5 6 ?)

	Parent trace 1	Parent trace 2	Recombined trace	Old repair	New repair
1	[1,2,3,4,5,6,7] : 1	[1,2,3,4,5,6,7] : 7	[1,2,3,4,5,6,7] : 1	[1,2,3,4,5,6,7] : 1	[1,2,3,4,5,6,7] : 1
2	[2,3,4,5,6,7] : 2	[1,2,3,4,5,6] : 6	[2,3,4,5,6,7] : 2	[2,3,4,5,6,7] : 2	[2,3,4,5,6,7] : 2
3	[3,4,5,6,7] : 3	[1,2,3,4,5] : 5	[1,2,3,4,5] : 5	[3,4,5,6,7] : ?	[3,4,5,6,7] : 5*
4	[4,5,6,7] : 4	[1,2,3,4] : 4	[1,2,3,4] : 4	[3,4,5,6,7] : ?	[3,4,6,7] : 4*
5	[5,6,7] : 5	[1,2,3] : 3	[1,2,3] : 3	[3,4,5,6,7] : ?	[3,6,7] : 3*
6	[6,7] : 6	[1,2] : 2	[1,2] : 2	[3,4,5,6,7] : ?	[6,7] : ?
7	[7] : 7	[1] : 1	[1] : 1	[3,4,5,6,7] : ?	[6,7] : ?
	Phenotype	Phenotype		Phenotype	Phenotype
	(1 2 3 4 5 6 7)	(7 6 5 4 3 2 1)		(1 2 ? ? ? ? ?)	(1 2 5 4 3 ? ?)

entries of the trace as the arguments of these types depend on the value that was changed. So, the overall effect of point mutation together with the old repair is effectively a macro-mutation, which is clearly an unintended effect of point mutation.

Using the new repair method, a single change in a value of the trace (point mutation) has only a limited effect on the subsequent entries of the trace, as the arguments of these types even if different are still compatible with the values down the line in the trace. So, in this case the overall effect of the point mutation together with the new repair is a much less disruptive form of mutation.

The exact effects of the old and new repair methods are dependent on the specific generator, but in general the new method is guaranteed to be less disruptive than the old.

In Table 3, we compare the disruption of the old and new repair methods on an illustrative example, that of iteratively sampling from a list to produce a permutation. Here, ? indicates that a resampling event is required, while * indicates "recycling" as described above. As we can see, the new repair mechanism triggers fewer resampling events.

5   Experiments and Results

The goal of our experiments is to explore the performance of PTO with GRASP-like generators on a range of COP problems, the effect of the novel trace repair mechanism, and the effect of the α parameter. Our goal is not to achieve state-of-the-art results, and in fact we can expect the results to be similar to those achieved by GRASP itself with the same parameters.

Our code is available in Github[1] and a dedicated script for the following experiments is available. The PTO implementation is written in Python, and has been adapted for execution in PyPy[2] for an approximate $8\times$ speed-up relative to pure Python (our bottleneck is algorithmic rather than numerical, so PyPy is more effective than the numerical library Numpy).

5.1   Problems and Instances

We have chosen a mix of dataset-based COP problems for realism and synthetic problem instances (of sizes comparable to real-world datasets) for controlled investigation of scalability. The TSP, JSSP and Knapsack problem are well-known. We use the simplest, most canonical version in each case, and in particular we use the 1-dimensional Knapsack. In the Ordering problem of size n, a "toy problem", the solution is a permutation of size n and the goal is simply to assemble the permutation $(12\ldots n)$. The objective function penalises each entry x_i in the permutation by $|x_{i-1} - x_i + 1|$. Since PTO is phrased as a maximising algorithm, we define Ordering fitness as the negative of that summed penalty. Similarly, TSP fitness is defined as negative cost, and JSSP as negative makespan.

For Ordering, we use the instances of sizes 10, 20, 40, 80, 160, 320. For Knapsack, we generate random instances of the same sizes. For TSP, we have taken 6 instances from TSPLIB[3], named att48, berlin52, eil101, u159, a280, and rat575 (where the integer gives the size). For JSSP, we have taken the instances azb5-azb9 [11] and yn4 [12] from the OR Library[4] file jobshop1.txt, with sizes 10×10, 10×10, 20×15, 20×15, 20×15, 20×20. Although these instances are not large, they are commonly used in modern JSSP research [13].

5.2   Experimental Design

We use three solvers plugged-in to PTO: Random Search (RS), Hill-Climbing (HC), and an Evolutionary Algorithm (EA). In HC a move is accepted if better than or equal to the current point. In the EA, we use 0.5-truncation selection, a crossover rate of 1 and a mutation rate of 1 per individual. The budget is set to 20,000 evaluations for all experiments, as in previous work [1]. For the EA, PTO internally sets the number of generations = population size = $\sqrt{20000} = 141$. This is in keeping with the PTO philosophy of minimising the amount of user configuration required.

---

[1] https://github.com/program-trace-optimisation/PTO.
[2] https://pypy.org.
[3] http://comopt.ifi.uni-heidelberg.de/software/TSPLIB95/.
[4] http://people.brunel.ac.uk/~mastjjb/jeb/orlib/jobshopinfo.html.

The novel trace repair mechanism introduced in Sect. 4 is compared against the original PTO trace repair mechanism. We will refer to these as *Strong* reuse and *Weak* reuse respectively. We compare 5 α values, $\{0.0, 0.1, 0.5, 0.9, 1.0\}$.

We have a total of (4 problems) $\times$ (6 instances per problem) $\times$ (3 solvers) $\times$ (2 re-use approaches) $\times$ (5 α values) $\times$ (20 repetitions) = 14400 runs.

5.3   Results

Results are shown in Fig. 2. By inspection, we have four main results. (1) Strong reuse is better than Weak reuse. The original trace repair mechanism, giving Weak reuse, was too disruptive for the type of generator (constructive heuristic) used on these problems. (2) Table 4 shows also the mean time taken per run. As expected, the time taken increases with problem size. There is no important difference in time taken between the Weak and Strong re-use methods. A further result is not shown: increasing α tends to increase the time taken, since it gives a larger RCL. (3) the HC solver is better than either RS or EA. The use of HC over RS amounts to the difference between using a combined constructive heuristic/search metaheuristic (as in GRASP and PTO) and using a constructive heuristic alone (with equivalent effort). The advantage is clear-cut. However, an EA does not improve performance further. (4) The other main result is that on the Knapsack and JSSP problems, an intermediate level of greediness ($\alpha = 0.5$) tends to outperform highly greedy construction (small α) or highly randomised construction (large α). For the Ordering problem, a purely greedy approach ($\alpha = 0.0$) solves the problem perfectly. This is the expected result since Ordering is unimodal. It is more surprising that a greedy approach does well on TSP.

Results 1–3 are clear-cut and quite consistent across problems and instances. Result 4 is more problem-dependent, as expected. Perhaps the best setup for difficult problems is the HC solver with $\alpha = 0.5$. For this setup, we give results across all problems in Table 4.

5.4   Discussion

We have seen that performance of hill-climbing is better than the evolutionary algorithm, overall. Two main hypotheses can be suggested to explain this result.

(1) Perhaps our crossover operator is disruptive, i.e. despite the improved repair mechanism, it fails to retain enough good information from the parents and recombine it in a way that it remains "good". The PTO crossover operator is a uniform crossover on the dictionary of trace entries. This choice was made because a one-point crossover on trace entries is difficult to define for the general case. However, there are some possibilities to do so, or to do so for the special case of GRASP-like generators. These will be considered in future work.

(2) When hill-climbing does well, it suggests that the landscape is somewhat unimodal. Thus, we may ask: what is the effect on the fitness landscape of using "smart" constructive heuristics as PTO generators? Our speculative

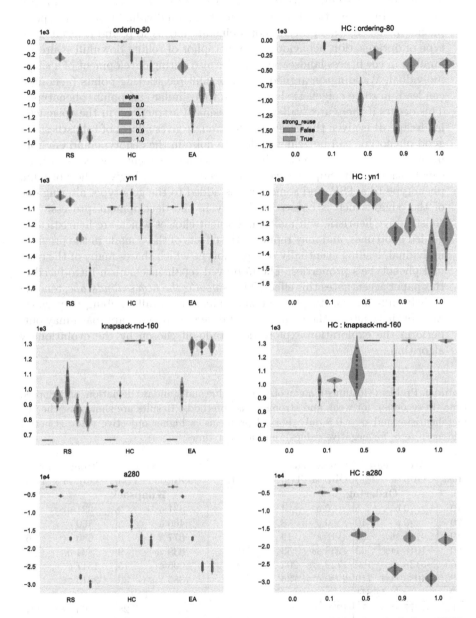

Fig. 2. Results for one instance of each problem type. Top to bottom: Ordering, JSSP, Knapsack, TSP. Left: strong re-use only; analysis by solver and α. Right: HC only; analysis by α and strong versus weak re-use. For all problems, higher values are better (closer to zero, for negative values). Horizontal axes (fitness) scaled by 10^3 or 10^4 as shown.

answer involves seeing constructive heuristics as epigenetic, developmental processes. In epigenetics, the development of an individual from genotype to phenotype is seen as a process, not an instantaneous step, and it has its own type of optimisation behaviour. The metaphor of rolling downhill – already familiar in the fitness landscape – applies also during development of a single individual. Waddington argued [14] that multiple starting points (genomes) can lead, in some epigenetic landscapes, to a similar end-point (phenotype). This occurs if there are "valleys" or "basins of attraction" in the epigenetic landscape. It is given the name *canalisation*. It gives a form of robustness – a good phenotype can be achieved despite noise in the genotype and environment. We can see constructive heuristics as developmental processes with a canalisation effect. Suppose we have a genotype which gives the optimum phenotype to a TSP, and some noise is added to the genotype, altering one of the edges which is constructed early in the development process. The constructive heuristic will make subsequent choices which are heuristically-guided good ones, and may replicate the effects of those made in the previous individual, leading eventually to a phenotype which shares many of the original phenotype's properties. The improved repair mechanism introduced in this paper accentuates this effect. This *smooths the fitness landscape*, tending to help hill-climbing to perform well. This allows hill-climbing to perform well. In this context, the highly exploitative hill-climbing approach may outperform the exploration-exploitation trade-off chosen by the evolutionary algorithm.

Table 4. Problem (instance), mean objective value and standard deviation, and elapsed time (in seconds), for weak and strong re-use methods. Results are shown for the HC metaheuristic and $\alpha = 0.5$ only. For all problems, a higher objective value is better (closer to zero, for negative values), and a lower time.

Inst.	Weak reuse Obj. sd	T(s)	Strong reuse Obj. sd	T(s)	Inst.	Weak reuse Obj. sd	T(s)	Strong reuse Obj. sd	T(s)
	Ordering					**Knapsack**			
10	0 0	5	0 0	4	10	331 0	3	331 0	3
20	-5 6	10	0 0	8	20	460 0	3	460 0	4
40	-175 26	22	-18 8	19	40	677 4	6	680 0	6
80	-1018 110	63	-238 36	53	80	839 45	9	894 0	11
160	-5138 405	190	-1722 178	203	160	1096 81	15	1317 0	23
320	-24088 1444	719	-9020 670	736	320	1385 146	30	2015 7	59
	JSSP					**TSP**			
abz5	-1277 21	77	-1270 19	81	att48	-64720 3891	28	-46665 2380	24
abz6	-1008 19	78	-1000 19	77	berlin52	-15564 790	32	-11063 477	27
abz7	-774 12	322	-783 15	325	eil101	-1861 56	88	-1263 79	78
abz8	-795 19	324	-793 18	328	u159	-206070 4381	186	-136750 7680	167
abz9	-817 13	323	-817 11	317	a280	-16806 464	546	-12612 610	506
yn1	-1046 20	487	-1043 20	469	rat575	-61021 1070	2571	-49457 1187	2448

6   Conclusions and Future Work

In this paper we have explored the use of "smart", GRASP-like, constructive heuristics as generators in the PTO framework, to solve combinatorial optimisation problems. They are a natural fit, and can be seen as a generalisation of previous work, or in a sense as a replication but with automation. We have introduced an improved PTO trace repair mechanism which runs after each genetic operation, and which gives stronger re-use of genetic material relative to the original. After extensive experimentation, we then have four main results:

1. The novel strong re-use method beats weak re-use;
2. Strong re-use does not incur a penalty in computation time;
3. HC beats RS and EA;
4. Medium levels of greediness in the heuristic are often best for real problems.

There are then many lines of research open for the future. Although for convenience in this paper we have introduced a GRASP-like common format for PTO generators for COP problems, generators going beyond this format are also possible. In fact, the freedom of the user to write or supply a generator in any format is a claimed strength of PTO, and in future work novel generators which do not emulate GRASP in this way will be introduced.

If we see the GRASP restricted candidates list as a stepped-uniform distribution on the remaining items ordered by their costs, then we can consider generalising by plugging in a different distribution, such as the triangular distribution suggested by Juan et al. [4]. There is also the possibility of using α to give a threshold on *rank of cost* as opposed to a threshold on *cost*.

The current PTO crossover is a uniform crossover, but a one-point crossover can also be defined, either for the special case of GRASP or for the general case.

Acknowledgements. Thanks to Carlos Fonseca for discussion and to anonymous reviewers. This work was carried out while JMcD was at University College Dublin.

References

1. Moraglio, A., McDermott, J.: Program trace optimization. In: Auger, A., Fonseca, C.M., Lourenço, N., Machado, P., Paquete, L., Whitley, D. (eds.) PPSN 2018, Part II. LNCS, vol. 11102, pp. 334–346. Springer, Cham (2018). https://doi.org/10.1007/978-3-319-99259-4_27
2. Feo, T.A., Resende, M.G.: Greedy randomized adaptive search procedures. J. Global Optim. **6**(2), 109–133 (1995)
3. McDermott, J., Carroll, P.: Program optimisation with dependency injection. In: Krawiec, K., Moraglio, A., Hu, T., Etaner-Uyar, A.Ş., Hu, B. (eds.) EuroGP 2013. LNCS, vol. 7831, pp. 133–144. Springer, Heidelberg (2013). https://doi.org/10.1007/978-3-642-37207-0_12
4. Juan, A.A., Faulin, J., Ferrer, A., Lourenço, H.R., Barrios, B.: MIRHA: multi-start biased randomization of heuristics with adaptive local search for solving non-smooth routing problems. TOP **21**(1), 109–132 (2013)

5. de Armas, J., Keenan, P., Juan, A.A., McGarraghy, S.: Solving large-scale time capacitated arc routing problems: from real-time heuristics to metaheuristics. Ann. Oper. Res. **273**(1–2), 135–162 (2019)
6. Ahuja, R.K., Orlin, J.B., Tiwari, A.: A greedy genetic algorithm for the quadratic assignment problem. Comput. Oper. Res. **27**(10), 917–934 (2000)
7. Bean, J.C.: Genetic algorithms and random keys for sequencing and optimization. ORSA J. Comput. **6**(2), 154–160 (1994)
8. Wingate, D., Stuhlmueller, A., Goodman, N.: Lightweight implementations of probabilistic programming languages via transformational compilation. In: Gordon, G., et al. (eds.) AISTATS. PMLR, vol. 15, pp. 770–778, 11–13 April 2011
9. Moraglio, A.: Towards a geometric unification of evolutionary algorithms. Ph.D. thesis, University of Essex (2008)
10. Jones, T.: Evolutionary Algorithms, Fitness Landscapes and Search. Ph.D. thesis, University of New Mexico, Albuquerque (1995)
11. Adams, J., Balas, E., Zawack, D.: The shifting bottleneck procedure for job shop scheduling. Manage. Sci. **34**, 391–401 (1988)
12. Yamada, T., Nakano, R.: A genetic algorithm applicable tolarge-scale job-shop instances. In: Manner, R., Manderick, B. (eds.) Parallel Problem Solving from Nature, vol. 2, pp. 281–290. North-Holland, Amsterdam (1992)
13. Bierwirth, C., Kuhpfahl, J.: Extended GRASP for the job shop scheduling problem with total weighted tardiness objective. Eur. J. Oper. Res. **261**(3), 835–848 (2017)
14. Waddington, C.H.: Canalization of development and the inheritance of acquired characters. Nature **150**(3811), 563 (1942)

Correction to: Evolutionary Computation in Combinatorial Optimization

Arnaud Liefooghe and Luís Paquete

Correction to:
A. Liefooghe and L. Paquete (Eds.): *Evolutionary Computation in Combinatorial Optimization*, **LNCS 11452,**
https://doi.org/10.1007/978-3-030-16711-0

1. Chapter 6.
In the originally published version the indexes of some variables in Section 4.1, including Constraints (4) and (5) of the model, include a wrong offset of one position. Some errors occurred in notations of variable indexes in Constraints (4) and (5) conditions of the model in Section 4.1, together with some ambiguities that may lead to misunderstanding for the reader. This was corrected in the updated version.

2. Chapter 11.
The originally published version of the paper "Clarifying the Difference in Local Optima Network Sampling Algorithms" contained an error. The additional text correcting the error has been added below.

Summary
During a re-analysis of the data-set, the lead author realised that she had made analytical errors while computing the results for this paper. This erratum presents the corrected numeric results in Section 3. These correspond to Tables 3 and 4 and Figures 2 and 3 — all of which are from Section 4.2 in the original paper. This report discusses the affect on the main conclusions of the work in the next Section. We found that while the numeric results are changed, most of the conclusions are still generally correct.

Affect on Conclusions
Conclusion 1. We found that the two sampling methods exhibited some agreement in the networks they produced and that we could reject the null hypothesis that they produce completely independent samples. They differed from a descriptive perspective in that *walkSample* was tuneable and predictable, while *optSample* varied widely but

---

The updated version of these chapter can be found at
https://doi.org/10.1007/978-3-030-16711-0_6
https://doi.org/10.1007/978-3-030-16711-0_11

seemed good at finding hub-and-spoke structure in the local optima space. **This conclusion is still correct.**

Conclusion 2. The correlations were stronger and clear when considering the features of the LONs obtained using *optSample* than *walkSample*. **This conclusion is still correct.**

Conclusion 3. We also worked on explaining heuristic algorithm performance on the problems using linear and random forest models, and found that the sampled LON features (for both *optSample* and *walkSample*) better fit the ILS response variable than the TS one. **This conclusion is now reversed**, i.e., the sampled LON features better fit the TS response variable than the ILS one.

Conclusion 4. We saw that generally, including both the funnel metric set and the network set would be advantageous in explaining search discrepancies for these two heuristics. **This conclusion is still correct.**

Conclusion 5. For both *optSample* and *walkSample*, the extracted funnel metrics proved useful. **This conclusion is still correct with a nuance**, i.e. the extracted funnel metrics proved useful with respect to TS as a response variable.

Conclusion 6. Going off the random forest models alone, *optSample* uniformly had more predictive power than its competitor, for these choices of instances and heuristics. **This conclusion is still correct overall**; *optSample* generally had more predictive power, but not uniformly.

Conclusion 7. From the random forest rankings, the most important predictors were those pertaining to fitness in the sampled networks: the fitness of funnel bottoms, and of nodes in general in the network. This hints that perhaps fitness levels in the local optima space are more pertinent to heuristic search than the subset of transition edges sampled by the LON algorithms. **This conclusion is still correct.**

This work is supported by the UK's Engineering and Physical Sciences Research Council (grant number EP/J017515/1). Data generated during this research are available from the Stirling Online Repository for Research Data (http://hdl.handle.net/11667/128).

Corrected Results

Table 1. Corrected results for Table 3 in the original paper. R^2 values for linear and random forest models to explain heuristic performance variation on the combinatorial problems.

Sample	Model	Features	Param.	ILS	TS
optSample	lm	all	set 1	0.124	0.963
optSample	lm	all	set 2	0.135	0.963
optSample	lm	all	set 3	0.182	0.964
optSample	lm	all	set 4	0.109	0.974
optSample	lm	funnel	set 1	0.114	0.125
optSample	lm	funnel	set 2	0.117	0.124

(continued)

Table 1. (*continued*)

Sample	Model	Features	Param.	ILS	TS
optSample	lm	funnel	set 3	0.162	0.127
optSample	lm	funnel	set 4	0.094	0.125
optSample	lm	network	set 1	0.069	0.957
optSample	lm	network	set 2	0.065	0.957
optSample	lm	network	set 3	0.075	0.958
optSample	lm	network	set 4	0.061	0.954
optSample	rf	all	set 4	0.307	0.820
optSample	rf	funnel	set 4	0.216	0.892
optSample	rf	network	set 4	0.247	0.775
walkSample	lm	all	set 1	0.640	0.980
walkSample	lm	all	set 2	0.339	0.955
walkSample	lm	all	set 3	0.294	0.964
walkSample	lm	all	set 4	0.158	0.968
walkSample	lm	funnel	set 1	0.200	0.956
walkSample	lm	funnel	set 2	0.104	0.946
walkSample	lm	funnel	set 3	0.086	0.954
walkSample	lm	funnel	set 4	0.061	0.901
walkSample	lm	network	set 1	0.062	0.027
walkSample	lm	network	set 2	0.245	0.042
walkSample	lm	network	set 3	0.248	0.117
walkSample	lm	network	set 4	0.112	0.016
walkSample	rf	all	set 4	0.045	0.910
walkSample	rf	funnel	set 4	−0.017	0.632
walkSample	rf	network	set 4	−0.084	0.249

Table 2. Corrected results for Table 4 in the original paper. Predictor rankings for the random forest models.

Sample	Features	Param	Resp.	1	2	3	4
optSample	funnel	set 4	ILS	sinkfitness	substrength	funnel	ncoptima
optSample	network	set 4	ILS	meanfitness	edges	outdegree	diameter
optSample	all	set 4	ILS	meanfitness	sinkfitness	ncoptima	funnel
walkSample	funnel	set 4	ILS	sinkfitness	ncoptima	funnel	substrength
walkSample	network	set 4	ILS	meanfitness	outdegree	edges	diameter
walkSample	all	set 4	ILS	outdegree	meanfitness	sinkfitness	ncoptima
optSample	funnel	set 4	TS	sinkfitness	ncoptima	funnel	substrength
optSample	network	set 4	TS	meanfitness	edges	diameter	outdegree
optSample	all	set 4	TS	sinkfitness	meanfitness	substrength	funnel
walkSample	funnel	set 4	TS	substrength	funnel	sinkfitness	ncoptima
walkSample	network	set 4	TS	meanfitness	outdegree	diameter	edges
walkSample	all	set 4	TS	sinkfitness	meanfitness	outdegree	ncoptima

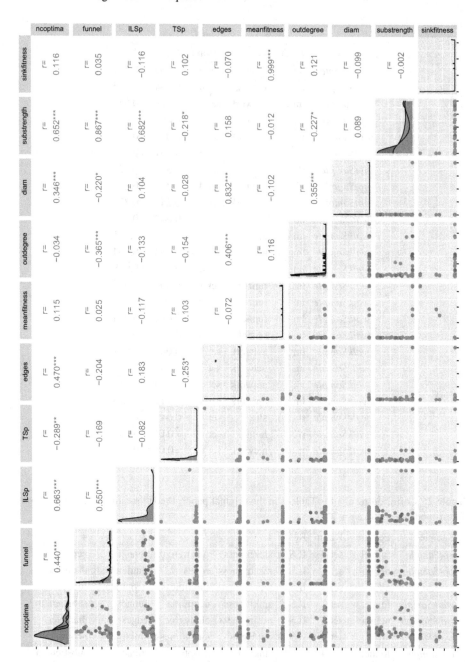

Fig. 1. Corrected results for Figure 2 in the original paper. Correlation matrix of performance metrics and *optSample*-produced LON features. Lower triangle: pairwise scatter plots. Diagonal: density plots. Upper triangle: pairwise Spearman's rank correlation, ***$p < 0.001$, **$p < 0.01$, *$p < 0.05$.

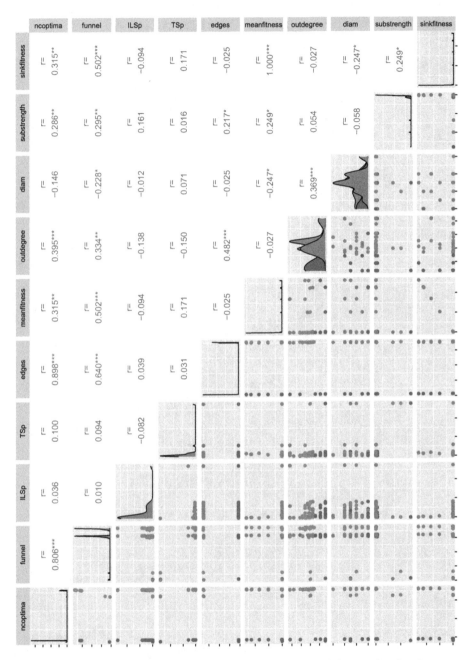

Fig. 2. Corrected results for Figure 3 in the original paper. Correlation matrix of performance metrics and *walkSample*-produced LON features. Lower triangle: pairwise scatter plots. Diagonal: density plots. Upper triangle: pairwise Spearman's rank correlation, ***$p < 0.001$, **$p < 0.01$, *$p < 0.05$.

Author Index

Printed in the United States
by Baker & Taylor Publisher Services